CONTINUOUS FLOW ANALYSIS

THEORY AND PRACTICE

CLINICAL AND BIOCHEMICAL ANALYSIS

A series of monographs and textbooks

EDITOR

Morton K. Schwartz

Chairman, Department of Biochemistry
Memorial Sloan-Kettering Cancer Center
New York, New York

ADDITIONAL VOLUMES IN PREPARATION

CONTINUOUS FLOW ANALYSIS

THEORY AND PRACTICE

William B. Furman

National Center for Drug Analysis
Food and Drug Administration
St. Louis, Missouri

with a chapter on THEORETICAL ASPECTS by

W. H. C. Walker

McMaster University Medical Centre
Hamilton, Ontario

CRC Press
Taylor & Francis Group
Boca Raton London New York

CRC Press is an imprint of the
Taylor & Francis Group, an **informa** business

First published 1976 by Marcel Dekker, Inc.

Published 2019 by CRC Press
Taylor & Francis Group
6000 Broken Sound Parkway NW, Suite 300
Boca Raton, FL 33487-2742

© 1976 by Taylor & Francis Group, LLC
CRC Press is an imprint of Taylor & Francis Group, an Informa business

First issued in paperback 2019

No claim to original U.S. Government works

ISBN 13: 978-0-367-45209-4 (pbk)
ISBN 13: 978-0-8247-6320-6 (hbk)

Visit the Taylor & Francis Web site at
http://www.taylorandfrancis.com

and the CRC Press Web site at
http://www.crcpress.com

LIBRARY OF CONGRESS CATALOG CARD NUMBER: 75-9088

INTRODUCTION

The explosive growth of clinical chemistry and analytical biochemistry during the past 20 years has been directly related to progress and development of automated techniques. The great majority of automated procedures are continuous flow methods, the automated system invented by Leonard Skeggs. Clinical, pharmaceutical, environmental, and industrial laboratories have all profited from these developments and indeed their present operation would be impossible without automation by continuous flow. The advent of continuous flow not only introduced the chemist to a new automated system, but also to a new form of chemistry and a new vocabulary to express the kinetics and dynamics of reactions in a continually flowing, but air-segmented fluid stream.

A difficulty in preparing a book of this kind is the enormity of the literature related to continuous flow. The 1975 Technicon Supplemental Bibliography, which includes a survey of the literature from 1957 through 1974, lists 7,870 references. Mr. Furman has successfully surmounted this problem and provides the user of continuous flow methods an insight into the theory of this form of chemistry and detailed descriptions of techniques for the assay of a variety of constituents.

Morton K. Schwartz, Ph.D.

PREFACE

Chapters 1 through 6 of this book comprise a literature search for techniques and notes on the operation of continuous flow analyzers, as typified by the Technicon AutoAnalyzer. About 3000 papers were reviewed during the search; if we assume that about 5000 papers have appeared in this area, the search represents a key to about 60% of the literature available through the middle of 1973. For the most part, the literature is reviewed noncritically. In fact, the careful reader will note several topics on which opinions from different laboratories are in direct conflict.

The orientation of the first six chapters is almost totally toward problems and solutions, not specific analytical procedures. No attempt has been made, for example, to list all the various automated methods of analysis for aspirin or for alkaline phosphatase. The chemistry of a system is described only when it helps explain why a certain technique was invoked. Many novel and fascinating chemical methods were therefore passed over, often with great reluctance. Articles in the rather specialized areas of amino acid analysis, air pollution analysis, or cell counters in blood analysis have been omitted unless the techniques described therein seemed useful to AutoAnalyzer users in general.

This book is intended mainly for laboratory personnel already having some experience with the AutoAnalyzer. Chapter 1 contains references to papers and books that provide a basic introduction to the AutoAnalyzer.

The casual reader interested in locating information on a narrow, specific topic (such as the permeability of tubing to carbon dioxide or an electrical checkout procedure for the Technicon Colorimeter-Recorder system) is advised to use the index as a guide.

The reader who wishes to gain a general background on the AutoAnalyzer is encouraged to begin with Chapter 2, reading those portions pertaining to equipment available in the reader's laboratory and stopping in each case at the section on special applications. Next, Chapter 7 should be read, for its contents will give great insight into the theoretical aspects of the AutoAnalyzer. Those portions of Chapter 4 that deal with techniques used in the reader's laboratory (extractions, distillations, etc.) should then be reviewed and, if the reader's interest goes beyond the standard Technicon modules, Chapter 3 may be consulted for information on the use of equipment made by other firms in continuous flow analytical systems. Chapters 5 and 6 are intended mainly for those involved with semiautomated column chromatographic separations and with computerized AutoAnalyzer systems.

After much thought, the decision was made to use a mixture of metric and English units. This book is meant to be a practical reference. The conversion of pump tube internal diameters from inches to millimeters will not be useful when one consults a catalog to purchase them. Moreover, suggesting that one consider the use of 101.325 to 202.65 kPa of air pressure on the inlet of an air pump tube simply generates a haze; 1 to 2 atm or 15 to 30 psi is, at present, more immediately understandable.

In reading the literature from the middle 1960s onward, it is often difficult to determine exactly which Technicon modules were used in a given method (e. g., Fluorometer I or Fluorometer II, etc.). The model numbers have been given whenever possible. In addition, some authors simply mention certain techniques or modifications of the equipment, without giving exact details. Such papers have been referenced nonetheless, with the thought that an experimenter may then get directly in touch with the authors of these papers for further information.

One of the greatest pleasures in the preparation of this book was the agreement of Dr. W. H. C. Walker to contribute the chapter on theoretical aspects. Both Dr. Walker and I wish to thank Dr. Ralph E. Thiers for reading Chapter 7 and providing valued suggestions and criticisms. I am grateful to Dr. Gerald Kessler, Mr. W. H. C. Shaw, and Miss I. Fortune for allowing excerpts from their papers to be directly quoted and to the C. V. Mosby Company and the Society for Analytical Chemistry for granting permission to reproduce these passages. Acknowledgments for permission to reproduce figures appear with the illustrations in the text. I also wish to thank Dr. Thomas P. Layloff, Professor of Analytical Chemistry, St. Louis University, for critically reviewing the entire manuscript; John C. Black, Food and Drug Administration, for redrawing several of the figures; James W. Myrick, Food and Drug Administration, for preparing the photographs in Chapters 2 through 4; and my wife, without whose constant encouragement and help this project could not have been realized.

Throughout Chapters 1 through 6, such words as Colorimeter, Fluorometer, and Recorder have been capitalized whenever referring to Technicon AutoAnalyzer modules. Acidflex, AutoAnalyzer, AutoAnalyzer II, Autograd, AutoTyper, AutoZyme, Coagulab, SMA 12, SMA 12/micro, SMA 12/30, SMA 12/60, SMAC, SOLIDprep, Solvaflex, and Technicon IDee are trade names of the Technicon Instruments Corporation.

William B. Furman
St. Louis, Missouri

CONTENTS

CONTENTS

CONTINUOUS FLOW ANALYSIS

THEORY AND PRACTICE

Chapter 1

CONTINUOUS FLOW ANALYSIS

I. INTRODUCTION

Taken in its broadest context, continuous flow analysis refers to any process in which the concentrations of single or multiple analytes are measured, uninterruptedly, in a moving stream of gases, liquids, or solids. In this sense, the concept has been in use for many years, notably in conjunction with chromatographic separations and manufacturing plant process monitoring systems.

Early in the 1950s, Dr. Leonard T. Skeggs, Jr., decided to apply continuous flow analysis to an ever-increasing workload of urine and blood samples in a large clinical laboratory. In collaboration with the Technicon Instrument Company, he developed a modular, automatic instrument, the AutoAnalyzer, which seemed applicable to many of the clinical tests then in use. The results of this work were reported in 1956 (772) and published in 1957 (767).

The AutoAnalyzer proved to be an extremely versatile instrument; by answering the need for automation of repetitive chemical tests, it had an impact not only in thousands of clinical laboratories throughout the world, but in government, academic, and industrial testing facilities. As a result, over 5000 papers reporting the use of the AutoAnalyzer have appeared in the years following its introduction.

By far, the most commonly used laboratory instrument for continuous flow analysis is the Technicon AutoAnalyzer, and it is this specific system, or application of the technique, that is discussed in this book.

II. DESCRIPTIONS OF THE AUTOANALYZER

Excellent descriptions of the early version of the AutoAnalyzer have been given by Skeggs (767, 773), Muller (615), Hainline (363), and Marsh (570). Comprehensive articles on the AutoAnalyzer in clinical chemistry have been written by Kessler (474, 475) and Lingeman and Musser (541). Whitby's discussion (900) of the pros and cons of automation in clinical chemistry, published in 1964, is still pertinent to a laboratory considering automation. Two general articles discussing the AutoAnalyzer in hospitals have appeared (35, 47), the latter treating the application of computers interfaced to the AutoAnalyzer. Zajac (931) shows a picture of a laboratory bench well designed for several AutoAnalyzers operated side by side and gives a schedule of preventative maintenance for most of the instrument's modules. White et al. (674) have written a book, treating almost all the practical aspects of the AutoAnalyzer, which any laboratory so equipped will find of great use. Ingram et al. (425) describe the problems involved in setting up an AutoAnalyzer in a small mobile van. Bernhard et al. (76) used an AutoAnalyzer system to analyze ocean water. To permit automatic analysis close to the point of sampling, an underwater container was designed for the AutoAnalyzer. The analytical signals were fed through cables up to the Recorders on board the ship. The submerged AutoAnalyzer system could be standardized whenever desired via remote control.

Brandstein et al. (130) and Moreland (605) discuss the all too often neglected hazards of operating AutoAnalyzers in poorly ventilated laboratory areas. Brandstein et al. (130) found that an AutoAnalyzer method for glucose, which used potassium cyanide and potassium ferrocyanide reagents, gave off hydrocyanic acid fumes at measured atmospheric concentrations exceeding the generally accepted safety limits. They give precautionary safeguard measures.

By the early 1960s, the single-channel AutoAnalyzer had become well accepted by many clinical laboratories, and the next logical step was to devise a multichannel, or sequential, arrangement whereby several analyses could be performed nearly simultaneously on a single test specimen. Levine (531), Skeggs and Hochstrasser (775), and Thiers et al. (831) describe early versions of multiple-channel AutoAnalyzer systems, and Whitehead has presented two lucid explanations of such systems (904, 906). Chasson et al. (179) evaluate a prototype twelve-channel sequential multiple analyzer (Technicon SMA 12), giving preventative maintenance tips and ideas on how to adjust and troubleshoot the instrument. Smythe et al. (781) discuss the theory of such an instrument, operated at 60 samples per hour (Technicon SMA 12/60), in which storage oscilloscope traces are used to diagnose and correct instrument malfunctions. Skeggs (769) has written a very readable history of the development of the AutoAnalyzer during the years 1957 to 1966.

III. SPECIAL AUTOMATED SYSTEMS

Red blood cell counts were performed on an AutoAnalyzer system described by Sturgeon and McQuiston (812). Particles or cells deflected a microscopic beam of light and thereby activated a photocell counting system. Drawings and an explanation of the Technicon cell counting detection system are included in their paper. Schultze (743) presented a somewhat related system, the Technicon Optical Somatic Cell Counter. This device used a dark field (forward scattering) particle counter system. The pulses caused by the passage of cells through the particle detector were displayed on an oscilloscope, and the counter threshold could be adjusted to eliminate pulse counts arising from particles of less than the selected minimum volume.

An automatic apparatus for enzyme rate determinations, the Technicon AutoZyme, is discussed by Morgenstern et al. (607). This apparatus incorporates a two-speed Pump with a speed ratio of 1:6. Operating at its slow speed, the Pump draws the enzyme sample, substrate, and reagents into a coil. The reaction proceeds for a predetermined period of time and yields a steady-state value on the Recorder. The Pump is then switched to high speed, moving the sample, substrate, and reagents rapidly through the coil. The recorded trace moves to a new steady state representing the value resulting from the shorter incubation time. The slope of the line recorded between the two steady states is a measure of the rate of the reaction, and from this slope the change in absorbance per minute of incubation can be calculated. The paper of Morgenstern et al. (607) includes the necessary equations. The AutoZyme unit, however, performs the calculations electrically and displays the result on the chart. Samples could be processed at 20 to 30 per hour.

Barabas and Lea introduced a method of automated anodic dissolution to prepare solutions of metals for analysis by an AutoAnalyzer. In one application (60, 61), phosphorus was analyzed in copper rod samples. The sample rod and a graphite rod were immersed in an electrolytic solution, and a timed, regulated current was passed between them. The sample solution thus produced was then automatically delivered via a solenoid valve into a cup on a Sampler II, and the analysis proceeded using a standard AutoAnalyzer. In a second application of the anodic dissolution apparatus

(59), steel rods were analyzed for manganese, nickel, and phosphorus using multiple AutoAnalyzer manifold channels.

The Technicon Coagulab system, used for blood coagulation profiles, was described by Adler and Tse (5). This is one of the very few "discrete" analyzer systems mentioned in this book. It is included because of its ingenious design principles and because it is a good example of a test well suited for discrete automatic analysis. Samples are transported through the apparatus by means of a strip of Mylar film. The film has rings preformed in it to hold the samples, and these rings contain a predeposited spot of magnetic iron oxide. At the first station, a drop of blood sample is automatically deposited on top of the iron oxide spot. The film strip is pulled to the next station, where a drop of thromboplastin reagent is added. The mixture is then moved to the next station, where a drop of calcium chloride reagent is added. Simultaneously, a timer is started. The sample is then moved under an optical "end-point detector," where a beam of light is reflected through the sample by a mirror to a phototransistor. The sample spot is spun through the action of a magnetic stirring bar, and at the beginning of the measurement the spot is opaque to the light beam. When coagulation occurs, the polymerized fibrin strands collect the iron oxide particles into clumps. The absorbance of the sample spot decreases sharply, causing the phototransistor circuit to stop the timer. The resulting prothrombin time is printed out by the system. "Drop detectors" are used to monitor the placement of the sample drop and reagent drops on the film carrier.

IV. DEVELOPING AUTOMATED METHODS

Of the many fine early reports in the literature, two papers are of particular interest in conveying the essence of method development for the AutoAnalyzer. In 1959, Ferrari et al. (273) presented a method for the antibiotics streptomycin and penicillin; this article gives both a detailed account of the equipment and the authors' "philosophy" of continuous flow analysis. In 1960, Lundgren (554) published a paper on phosphate analysis which amounts to a case history of how one approaches the job of adapting a classical colorimetric procedure to the AutoAnalyzer.

There are several useful articles covering general operational hints and problem-solving techniques for the AutoAnalyzer. Papers by Kessler (473-475), Lingeman and Musser (541), Marsh (570), and Skeggs (773) contain many helpful ideas and discuss "pit falls" to be avoided. Specific approaches used to adapt manual methods to the AutoAnalyzer are given by Marsh (570) and Schwartz (746), as well as in the Technicon instruction manuals (46, 680). Gerke and Ferrari (323) discuss problems and advantages in automating chemical and microbiological assays for antibiotics, and Jones (442) reviews solutions to problems encountered with automatic peptide analyzers. Whitby et al. (901) mention the various sources of errors from AutoAnalyzers in clinical chemistry, and Nisbet and Simpson (627) have presented a comprehensive study of drift errors found in various automated clinical tests.

Eiduson and Hoffman (238) set forth the requirements for acceptance of automated methods by the Association of Official Analytical Chemists (AOAC): proof of reliability, accuracy, precision, practicality, ready availability of instrumentation, and a successful collaborative study. Heuermann et al. (392) give a format for reporting automated methods to the AOAC and recommend how to conduct interlaboratory collaborative studies.

V. REVIEW ARTICLES AND BIBLIOGRAPHIES

Review articles provide the quickest access to the literature on automated analysis. Papers discussing the features and advantages of various automatic analysis instruments appeared in 1963 (106), 1969 (19), and 1970 (47). A very comprehensive review of automated techniques in pharmaceutical analysis was published in 1969 by Kuzel et al. (511) which covers most of the currently available automatic apparatuses. Two bibliographies have been prepared by Technicon, one covering the literature from 1957 to 1967 (826) and the other concentrating on later papers in the area of pharmaceutical analysis (827).

Table 1 lists nineteen review articles along with the general fields of interest covered by each. Blaedel and Laessig (104) give an extensive bibliography of "continuous" methods, covering the literature through 1963. In addition, they discuss the concept of response time and the dependence of response on sample size. They develop the idea that continuous flow systems are less sensitive than "discrete" analyzers, because the latter may be adjusted to carry nearly all the sample originally fed into the system through to the readout device, whereas a continuous flow system usually sends some of the sample to waste via stream splitting, debubbling, and so forth.

Evans and Thomas (248) discuss the various detection systems used with AutoAnalyzers as well as the techniques, e.g., steam distillation, available to the analyst using such equipment. Kessler gives operational hints and chemical explanations for several automated lipid determinations (473) and has written a particularly good article on the setup and operation of many clinical tests on the AutoAnalyzer (475). Schwartz (746) covers the calculations of enzyme activities using the AutoAnalyzer.

TABLE 1

Review Articles on Automated Analysis

Field of interest (year published)	Reference
Antibiotics (1968)	Gerke and Ferrari (323)
Clinical chemistry	
(1963)	Kessler (474)
(1966)	Lingeman and Musser (541)
(1970)	Kessler (475)
Enzyme activity	
(1963)	Schwartz and Bodansky (748)
(1968)	Schwartz (746)
Enzyme-catalyzed reactions (1964)	Blaedel and Hicks (101)
Enzyme kinetics (1968)	Schwartz and Bodansky (749)
Food analysis (1965)	Marten (575)
General bibliography (1966)	Blaedel and Laessig (104)
Kinetics in quantitative analysis (1968)	Pardue (645)
Lipid chemistry (1967)	Kessler (473)
Microbiological techniques (1964)	Ferrari (261)
Nutritional research (1966)	Evans and Thomas (248)
Peptide chromatography (1970)	Jones (442)
Pesticide residues (1966)	Gunther and Ott (355)
Pharmaceutical analysis (1969)	Kuzel et al. (511)
Protide analysis (1962)	Marsh (571)
Steroid assays (1968)	Russo-Alesi and Khoury (704)

Chapter 2

CONTINUOUS FLOW INSTRUMENTATION:
PRACTICAL CONSIDERATIONS
OF AUTOANALYZER MODULES

I. INTRODUCTION

Scattered throughout the literature of continuous flow analysis are hundreds of descriptions of problems, some common, others unique, encountered with the various AutoAnalyzer modules. These problems are usually solved during the course of research, leading to a successful automated method, and are often reported in papers as a sidelight to the main topic. Other authors discuss new uses or adaptations of the modules that are of interest to those using the AutoAnalyzer more as a research tool than as an analytical production instrument.

The purpose of this chapter is to compile and organize these bits of information and to allow ready reference to the source literature. Each of the common AutoAnalyzer modules is considered in turn. For each module, a section is presented on uses of the module itself, and this is followed by a section on specialized or advanced techniques.

II. COLORIMETER

A. Colorimeter-Recorder System

1. Descriptions and Notes on Operation

White et al. (674) discuss the operation and maintenance of the Colorimeter and Recorder, and mention that the Recorder amplifier gain should be checked with every change of Colorimeter filters. Thiers et al. (834) point out that if the Recorder gain is set too low, the curves show symptoms that mimic those of sample interaction.

The AutoAnalyzer Colorimeter and Recorder function electrically as one unit. The schematic of the original design is given by Skeggs (767). Kessler (474, 475) has written an excellent description of how the units function:*

> Light falling on the reference photocell generates a voltage across the slidewire of the recorder. The transmitted light falling on the sample photocell generates a voltage across the 100% transmission (T) potentiometer. The difference between the voltage developed across the 100% T control and the voltage across a portion of the slidewire is fed into the input of the chopper-amplifier in the recorder. The output of the amplifier powers the 2-phase balancing motor (which is coupled to the moveable arm of the slidewire and to the stylus) and rotates the arm in such a direction that the input to the chopper-amplifier is diminished. When the input to the amplifier is zero, the balancing motor comes to rest, and the position of the stylus is indicative of the ratio of the voltage developed on the sample side to the voltage developed on the reference side.

Owen (641) describes a simple test box circuit to check the AutoAnalyzer Colorimeter-Recorder system, giving both the schematic and clear, complete directions for its use.

Sawyer et al. (727) modified the Colorimeter-Recorder system, placing the baseline on the left of the recording with the peaks driving the Recorder pen to the right. This was done by reversing the L_1 and H_1 leads at the Recorder, and reversing the polarity of the two photocells in the Colorimeter.

2. Range Expansion

Taylor and Marsh (825) described what has become the most commonly used device to obtain scale expansion from the Colorimeter-Recorder system. A resistance selector box is placed in series between the negative side of the

---

*From G. Kessler, "An automated system of analysis," in Gradwohl's Clinical Laboratory Methods and Diagnosis (S. Frankel, S. Reitman and A. C. Sonnenwirth, eds.), 6th and 7th eds., Vol. I, C. V. Mosby Co., St. Louis, 1963 and 1970.

reference photocell and the corresponding Recorder input lead.* Usually, fixed value resistances are used (176, 493, 594) and a Heathkit resistance substitution box has also been so employed (41, 825). A variable resistance (potentiometer) gives additional flexibility (54).

Owen (642) gives a most complete description and schematic for such range expansion. As before, the resistance is placed in series between the reference photocell and the Recorder input (lead L on the Recorder). Variable or fixed resistors may be used, and a bypass switch is included to return the Colorimeter-Recorder system to an unmodified state. Owen describes how to calculate the required value(s) for the resistance, and we repeat his procedure and example:

First, one sets the system to 100% transmittance (T) with a blank. Second, one runs the highest standard and notes the peak reading as the Recorder pen moves to the left. Third, one decides how much farther to the left the pen should go and expresses this increment as a fraction of 100 less the original standard reading. Finally, one calculates the required resistance, in ohms, by multiplying the fraction obtained above times 1500. Thus, in Owen's example (642):

Blank reads 100% T

Highest standard reads 46% T

Desired highest standard reading is 10% T

Calculate: without expansion, 100 - 46 = 54

Additional displacement desired, 46 - 10 = 36

Fraction, 36/54 = 0.67

Resistance required, 0.67 x 1500 = 1000 Ω

Campbell (162) describes a convenient arrangement wherein the resistances are placed inside the Colorimeter itself. The lamp on-off switch is removed and replaced with a rotary switch that switches on the lamp and selects the desired resistor (i.e., the desired scale expansion factor). A schematic is given for this modification.

---

*By error, their schematic (825) shows the resistance box in series with the negative side of the sample photocell. It should be in series with the negative side of the reference photocell (private communication from Dr. Max M. Marsh, March 7, 1972).

Passen and Von Saleski (649) give a schematic of a range expander designed for use in inverse colorimetry, in which the transmittance increases with sample concentration. Their system yields expansion factors of 30, 50, and 70%.

Burns (152) and Thomas (837) found that very small apertures were required in the Colorimeter when scale expansion was used. Burns (152) reported that, when such small apertures were used, the Recorder damping had to be carefully adjusted to avoid sluggish response.

Although several authors (225, 784, 785) report good linearity when scale expansion up to 10x is used, others report that exact corresponding multiples are not always obtained (752), or that a system which gives linear response without scale expansion may become increasingly curvilinear as expansion is increased (837). Suffice it to say that linearity should always be verified, not assumed, when scale expansion is employed.

Equations are given for use with Recorder readings taken during scale expansion by Coote et al. (198) for "logarithmic units," by Ott and Gunther (638) and Pollard et al. (669) for obtaining corrected transmittance values, and by Scobell et al. (752) for calculating percent transmittance spans and absorbance values.

Gehrke et al. (315) note that, by substituting an 8-mm flowcell for a 15-mm flowcell and by using 2x range expansion, one obtains an approximately twofold logarithmic decrease in sensitivity plus a twofold electronic, linear increase in sensitivity. The result is an increased slope for the standard curve and a more nearly linear plot of concentration versus percent transmittance.

Expansion can be combined with a "chemical offset" to obtain scale expansion over a selected concentration range. For example, Blezard and Fifield (115) wished to expand the Recorder scale so that standard solutions representing 12 to 16% silicon dioxide would read nearly full scale. They placed an 11% SiO_2 standard solution in the Sampler wash box supply and adjusted the expansion to 4x. This combination of "chemical offset" plus expansion yielded a span of 80 recorder units for 12 to 16% SiO_2 standard solutions.

Range expansion may also be achieved without the use of added resistance between the Colorimeter and Recorder. Varley (869) set 100% T while aspirating the lowest standard solution (e.g., 10 ppm) and then calibrated the system by running the higher standards (e.g., 15 and 25 ppm). This offset the low end of the scale and tended to expand the range of interest on the Recorder. The degree of expansion was set with the Colorimeter zero potentiometer. Blanks were analyzed by adding the equivalent of 20 ppm standard to the blanks and then comparing them to the 20 ppm standard itself. Wagner (880) described an AutoAnalyzer method for which, in the normal mode, the Colorimeter output changed 10 chart divisions for a 1% change in standard concentration. When the proper aperture was inserted in the reference side of the Colorimeter, the Colorimeter output changed 25 chart divisions for the 1% change in standard concentration. This approach placed the 100% T line (the "baseline") off scale.

B. Colorimeter

1. Notes on Operation

Kessler (474, 475) mentions that the purpose of the "chimney" on the Colorimeter is to maintain the photocells and connections at acceptable temperatures by providing an outlet for heat.

Schunk (744) found that Colorimeter lamps burn out after 4 to 10 days of continuous operation. When a resistor was installed in series with the lamp so as to reduce its lumen output by about 50%, the lamp's lifetime was increased to 8 to 10 weeks of continuous use. This modification reduced the sensitivity of the Colorimeter-Recorder system somewhat, and the Recorder motor torque had to be carefully checked due to the reduced signal level.

It is sometimes necessary to reduce the reference beam energy below that afforded by the AutoAnalyzer Colorimeter's No. 1 aperture. This is often the case when working with inverse colorimetry or when the blank itself yields a fairly high absorbance. Devices used for this purpose include undersized apertures (343), partially masked No. 1 aperture plates (329, 330, 477), or a No. 1 aperture plus a piece of partially exposed photographic emulsion (298, 912) or a neutral density filter (329, 330, 477).

2. Nonlinearity

Although the interference filters supplied for the Colorimeter are nominally rated at a certain wavelength, it is good practice to obtain a complete scan of transmittance versus wavelength, using a recording spectrophotometer, to make sure of their optical characteristics (119). Certain filters may possess a second window (606) and, if present, light from the source lamp passing through the second window and the flowing solution may cause an additional response of its own, provided that the photocells are sensitive to energy at the wavelength of the second window. This additional unwanted response may cause marked deviation from linearity, especially when measuring high values of absorbance at the primary wavelength of the filter.

Often, these problems are most noticeable near the wavelength limits of the Colorimeter. Pentz (654) operated the Colorimeter at 352 nm and used Corning filters, in addition to the AutoAnalyzer interference filters, to reduce unwanted radiation. Wilson (913) used a Colorimeter equipped with "red-sensitive" photocells at 814 nm and employed "red-glass" filters to eliminate radiation from the other transmission bands of the interference filters. To achieve wavelength selectivity and low levels of extraneous light, Kessler and Wolfman (477) used special interference filters (Schott-Genussen DAL). Williams and Lyons (911) found that the AutoAnalyzer phototube Colorimeter was not linear past 0.4 absorbance at 340 nm and substituted a Beckman DB-G spectrophotometer.

In some cases, nonlinearity may be traced to the improper choice of a filter whose transmission band does not lie completely within the corresponding absorption band of the flowing solution (423, 606).

Nonlinear response from the Colorimeter may also be caused by extraneous radiation from laboratory lights, and linearity sometimes may be improved by placing nontransparent plastic tape around critical areas, such as the sample flowcell compartment cover (477, 654).

Morgan et al. (606) mention that the AutoAnalyzer Colorimeter may allow more scattered light to reach the photocell or phototube when the larger circular apertures are used and that substitution of smaller apertures may improve adherence to Beer's law.

Ryan and Morgenstern (711) reported that the use of Technicon's "black-glass" flowcell (15 mm, Technicon 170B058-1) helped maintain a linear response to intensely colored dye solutions at 505 nm.

3. Drift

Several approaches have been reported to reduce Colorimeter drift. White et al. (674) mention that the standard photocells show a "fatigue effect" and must be warmed up before they stabilize. Gehrke et al. (314, 318) placed the Colorimeter in an enclosure to shield it from drafts and sudden changes in ambient temperature. Goodall and Davies (335) matched Colorimeter photocell outputs by connecting the outputs from each to a potentiometer and tapping off a suitable fraction to galvanometers, one for each photocell. Sawyer et al. (727) found that drift could be minimized by selecting the optimum-sized reference aperture while aspirating the blank: Too large or too small an aperture gave drift in opposing directions on the Recorder.

4. Flowcells and Debubblers

a. Descriptions of Flowcells (263). White et al. (674) describe both the original (N-type) and the more recent tubular (S-type) flowcells. Ferretti and Hoffman (275) outline modifications of the N-type Colorimeter flowcell to improve its characteristics. Delves and Vinter (224) converted the N-type Colorimeter to the S-type, using Technicon's modification kit.

They then constructed a mask, fitted around the flowcell, to allow only the light passing through the flowcell to fall on the photocell detector.

b. Effects of Flow Rate and Debubbler on Carryover (Sample Interaction). Several authors report that faster flow rates through the tubular flowcell improve (reduce) the system's sample interaction or carryover (185, 248, 306, 356, 357, 884). In order to exceed a flow rate of 3.9 ml/min (the largest pump tube size), Christopherson (185) sent the flowcell exit stream to waste and pumped out the debubbler overflow stream. Gardanier and Spooner (306) recommend as high a flow rate through the flowcell as possible, discarding only the air bubbles plus a slight amount of liquid. They point out that, when the flowcell pull-through rate is slow, one actually observes a mixed average of several liquid segments entering the debubbler and that, when the rate is fast enough, one can almost monitor each individual segment.

The flowcell debubbler itself, through the mixing of the entering liquid segments, contributes heavily to sample interaction or carryover in the AutoAnalyzer system (356, 357, 884). Lane (518) modified the C5 debubbler (Fig. 1) by pushing a short section of 0.045-in. pump tubing up into the outlet arm, flush with the debubbler chamber, and then inserting one end of a 2.5-in. length of polyethylene tubing [(0.015 in. internal diameter (i.d.) x 0.043 in. outer diameter (o.d.)] into the section of pump tubing which acted as a sleeve. The other end of the 2.5-in. tube led the stream to the flowcell.* This modification reduced the internal volume of the C5 debubbler to a minimum, which improved the flow dynamics by reducing mixing in the debubbler. Christopherson (185) replaced the C5 debubbler with a C3 fitting. The capillary exit arms of the latter fittings improved sample resolution, compared to the large-bore exit arms on the C5s. Small-bore Teflon tubing (0.02 in. i.d.) was used between the debubbler and the flowcell. Teflon tubing, 24-gauge, was used to fill as much of the large-bore flowcell entrance arm as possible, and 18-gauge Teflon tubing was then inserted coaxially through the 24-gauge tubing to carry the liquid flow to the flowcell.

Debubblers may also be used purposefully as the final mixers in Auto-Analyzer systems. For example, Ott and Gunther (638) used two debubblers in series ahead of the Colorimeter flowcell to reduce background noise on the recorded trace.

c. Schlieren. The tubular flowcell Colorimeter may give responses even to essentially colorless solutions, due to changes in refraction of light or density in adjacent segments fed to the flowcell from the debubbler (908,

---

*By error, Lane's drawing (518) indicates a 5.5-in. length of tubing between debubbler and flowcell. The correct value is 2.5 in., as indicated in Fig. 1 (private communication from Jonathan R. Lane, April 25, 1973).

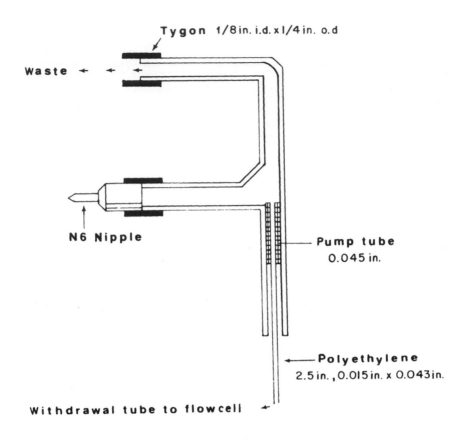

FIG. 1. Lane's modified C5 debubbler. The internal volume is minimized, thus reducing mixing and sample interaction. Reproduced from Lane (518) by permission of the author and the Association of Official Analytical Chemists.

909). These changes cause readily seen ripples, or streamers (schlieren), in the flowing solution and may occur in colored or colorless streams. The problem has been cured by better mixing (159) or by use of pulse suppressors plus substitution of a B4 fitting for the usual flowcell debubbler (543). The B4, with its larger volume, promoted thorough mixing of segments before they passed through the flowcell. Dow (230) reported that schlieren in viscous streams may be reduced by avoiding abrupt changes in flow rate as the solution is passed through an 8-mm tubular flowcell. This was effected by approximately matching the internal diameter of the inlet and outlet tubing leading to and from the flowcell. Gaddy (295) used the AutoAnalyzer to measure various inorganic substituents in heavy water.

Standards were dissolved in the expensive, pure heavy water rather than normal water to avoid "interfaces" in the flowcell due to density difference between heavy and normal water.

d. Foaming and Coating in Flowcells. Occasionally, foaming may occur in the flowcell. Failing et al. (254) solved the problem by placing a small amount of siliconized stopcock grease on the tip of the tubing leading into the flowcell. White et al. (674) mention lightly coating the inside of the S-type flowcell with silicone grease to eliminate foaming.

Flannery and Steckel (280) found that the use of 0.4% Brij-35 solution reduced the formation of a coating in the Colorimeter flowcell arising from the accumulation of a color complex in a determination of magnesium.

Bano and Crossland (58) used a Colorimeter to monitor a chloroform phase and found that they could conveniently cleanse the flowcell of traces of water by flushing it with about 5 ml of "methylated spirit" (5% methanol in ethanol) using a syringe. Water traces in the flowcell often cause a noisy recording in such systems.

e. Surging. "Surging" may occur when the reaction stream from a Heating Bath coil is fed to the Colorimeter flowcell. Lenard et al. (528) led the stream from the Heating Bath, through a jacketed cooling coil, to the Colorimeter debubbler and found that placing a Tygon pump tube (0.020 in. i.d.) between the jacketed coil and the debubbler prevented surging which otherwise forced air bubbles through the debubbler into the flowcell. Passen and Von Saleski (649) encountered surging from a Heating Bath coil which fed the solution to the Colorimeter debubbler. They reduced the surging and improved the recorded baseline by leading the solution from the flowcell through a 6-in. length of tubing into a flask and then pumping the solution out of the flask to waste. Tietz and Green (841) recommend channeling the stream from the Heating Bath through a sleeve tubing (0.125 in. x 0.1875 in. diameter) and slipping the tubing directly over the flowcell entrance, using no nipples or narrow tubing between the sleeve tubing and flowcell. The same authors (842) later reported a special arrangement of fittings to reduce surging. The stream from a Heating Bath was led into the lower entrance of the bulb portion of a 1-ml volumetric pipet from which the stems had been removed. The stream flowed upward through the bulb and then entered the side entrance of a B2 gas trap from which the capillary protruding into the trap had been broken off. From the bottom of the B2, the stream was led to the flowcell debubbler. The bulb and B2 fittings were placed in a vertical position, taped onto the side of the Colorimeter.

Another way to smooth the flow through the flowcell is to create a backpressure at the debubbler. This has been achieved by using a pulse suppressor (0.020 in. i.d.) (308) or a coil of capillary Teflon tubing (551) on the waste line outlet from the flowcell debubbler.

f. Comparisons of Flowcells. Powell et al. (673) compared the Acculab flowcell (1.5 mm x 15 mm) to the standard Technicon flowcell and found that the former gave increased sensitivity and improved washout characteristics.

Robertson et al. (687) used a manifold in which the final colored species was extracted from an aqueous phase into a chloroform phase. The concentration of the colored species was then measured by monitoring the absorbance or transmittance of the chloroform phase. Traces of water in the chloroform phase interfered with the measurement, but certain flowcell designs were found more susceptible to such interference than others. The interference observed using the Technicon 15-mm tubular flowcell was compared to that of several other flowcells. The Technicon flowcell performed satisfactorily if kept completely free from water.

5. Comparisons of Colorimeters and Spectrophotometers

Morgan et al. (606) discuss the merits of the phototube and photocell Colorimeters and conclude that the emissive, high-resistance phototube is more accurate than the low-resistance photocell.

Schwartz et al. (751) used a 1-cm pathlength flow cuvette in the Colorimeter and found that, at about 680 nm, the Colorimeter gave absorbances about 75% of those found when the same cuvette was used in a Beckman DU spectrophotometer, which had been modified to hold it. They ascribe this to the use of filters in the Colorimeter versus monochromatic light in the DU. Kessler et al. (476) found that the Colorimeter II and the SMA 12/60 Colorimeter gave absorbance readings at 340 nm which were 80% of those obtained from a Cary 14 spectrophotometer, taking pathlength corrections into account. These Colorimeters were linear to at least 0.6 absorbance.

C. System II Colorimeter

Smythe et al. (781) furnish a drawing of the System II Colorimeter flowcell. The design of the flowcell places the debubbler very close to the optical path, thus reducing dead volume and improving the wash.

Levy and Konig-Levy (537) discuss the problems involved when feeding solutions from a Dialyzer I (203-cm dialysis pathlength) to the Colorimeter II flowcell-debubbler. The ratio of liquid to air becomes critical: Too much air may defeat the debubbler and allow air to occasionally pass through the light beam; if there is too little air, the wash becomes poor. These authors recommend a ratio of about 5:1 of liquid to air (e.g., 1.2 ml/min liquid to 0.23 ml/min air) in the Dialyzer I recipient stream and a ratio of about 1.5:1 of liquid entering the flowcell debubbler to that passing through the flowcell (e.g., 1.2 ml/min to 0.8 ml/min).

D. Multiple-Cell Colorimeters

Levine (531) describes an early version of the AutoAnalyzer multiple-cell Colorimeter. Skeggs and Hochstrasser (775) used a Colorimeter which moved up to eight different flowcells sequentially into the sample light path via a rack driven by a motor pinion and described the timer systems used to control the cell rack position. Whitehead (904) gives a figure and description of an eleven-channel Colorimeter in which the eleven flowcells were intermittently positioned in the light path.

E. Differential Colorimeters

Russo-Alesi et al. (705) discuss a two-flowcell colorimeter, operable in "dual" or "difference" modes. In the former mode, the signals from the two flowcells are sent to separate recorders, and blank subtraction is done manually during the computations. In the latter mode, the colorimeter electrically subtracts the blank signal and sends only the one corrected sample signal to a recorder. The sample stream flows through one of the flowcells and the blank stream flows through the other; the filters in both beams must be well matched for good precision.

Hamilton (372) describes a split-beam, self-compensating colorimeter in which two flowcells are used. A time-delay coil was placed between the flowcells, so that the same stream was monitored twice, at different times. This reduced the apparent baseline drift. When the length of the delay coil was properly chosen, the recorded trace resembled a "differential" curve.

Jansen et al. (435) designed a two-channel, filter colorimeter in which one tungsten lamp provided source energy for two flowcells mounted 180 deg from each other with the lamp between them.

F. Special Applications

1. Manual Colorimetry

The Colorimeter-Recorder system may be used, independently of other AutoAnalyzer modules, for rapidly measuring the absorbance of manually prepared solutions. Grossman et al. (350) describe such an arrangement, which featured a variable chart speed mechanism and a switch between the Colorimeter photocell output and the Recorder. The prepared solutions were fed into the Colorimeter flowcell through a funnel.

2. Simultaneous Determination of Two Species

Garritsen et al. (308) analyzed simultaneously for copper and nickel by forming their ethylenediaminetetraacetic acid (EDTA) complexes, resulting in a mixture of two colored species. Two Colorimeters, set at the appropriate wavelengths, monitored the stream and, through the use of simultaneous equations, the two recorded absorbances yielded the concentrations of the metals. These authors also demonstrate how to construct a nomograph to obtain rapid, approximate results during on-stream plant operation. In a similar experiment, Hoyt and Jordan (416, 417) determined iron and aluminum on the same manifold using Colorimeter readouts at two different wavelengths corresponding to the differences in spectra of the metallic complexes formed with "Ferron" reagent.

3. Monitoring Wide Absorbance Ranges

Ferrari and MacDuff (272) used tandem Colorimeters, at different wavelengths, to achieve various sensitivity ranges while monitoring a flowing stream. Tappel and Beck (822) arranged four Colorimeters to monitor absorbances of greatly varying magnitudes. The manifold effluent was led through the first two Colorimeters equipped with filters at different wavelengths (420 and 505 nm). The outflow from the second Colorimeter was sent back to the manifold, where it was diluted, and then read through the third and fourth Colorimeters, set at the same two wavelengths.

Another solution to the problem of reading widely varying absorbances was given by Lenard et al. (528). Their AutoAnalyzer system monitored a sample stream, pumped continuously from a reaction vessel, to measure kinetic parameters. During the run, the absorbance occasionally became too high to read accurately. At such times, the change or removal of a Colorimeter aperture plate would bring the signal back into the usable range. In one example, the Colorimeter was balanced (zeroed) with a No. 2 aperture plate in the reference beam and a No. 1 aperture plate in the sample beam. When the absorbance became too high, removal of the sample aperture plate gave a constant decrease in reading of about 0.56 absorbance. This decrease was found to be independent of the absorbance reading before removing or substituting the sample aperture plate.

4. Turbidimetry

The Colorimeter has been used by several authors for monitoring solutions turbidimetrically. The wavelengths used span a rather wide range: 410, 420, 520, 550, 580, and 660 nm (460, 65, 227, 221, 549, and 619, respectively). Basson and Bohmer (65) used a 50-mm flowcell to measure sulfate by following the turbidity formed by addition of barium chloride reagent, and noted that high flow rates through the flowcell resulted in poorly shaped recorder peaks (shown in their article), which became acceptable at a lower flow rate (0.6 ml/min). Dieu (227) analyzed for sulfates by turbidimetry, using a barium chloride/gelatin reagent, and found that the flowcell became coated with barium sulfate. This problem was overcome by the use of "air rinsing," a technique found superior to other attempted remedies (e. g. , rinsing with solutions of EDTA or phosphoric acid in every other cup). A Technicon three-way valve was placed just ahead of the Colorimeter and, at timed intervals, this valve diverted the flow stream to waste and directed air through an A0 glass fitting into the flowcell debubbler for 22 sec. The three-way valve was controlled by the upper portion of a special Technicon double cam in the Sampler. The two cams were adjusted to flush the cell with air immediately after a peak was recorded.

5. Electronic Circuits

Hamilton (372) gives an excellent summary of electronic circuits associated with colorimeters in general, with many references to the literature. Signal amplification and logarithmic conversions are discussed.

a. Connection of Colorimeter to Standard Laboratory Recorders. Habig et al. (356, 357) give a circuit diagram to connect the Colorimeter sample photocell to a Heath EUW-20A recorder, so as to record signal changes which are too rapid for the standard Technicon Recorder to follow. Katz and Levy (454) give a schematic to connect a Heath EU-20B recorder to the Technicon Colorimeter. In their circuit, the recorder's internal reference voltage is disconnected, being replaced by the Colorimeter's reference photocell signal. The Colorimeter is adjusted, as usual, by its 0% T and 100% T potentiometers. The recorder worked well, except in regions of the spectrum where the Colorimeter developed low-level signals, such as at 410 nm.

b. Operational Amplifier Circuits. Operational amplifiers have been used to amplify and condition the signal from the Colorimeter. Lapidus et al. (519) published a schematic to transform the Technicon Colorimeter photocell signal into a signal proportional to concentration. Their circuit uses the "two-point" standardization technique. Thus, the standard is passed through the AutoAnalyzer manifold, and the desired recorder reading is obtained by adjusting a potentiometer in the circuit. The baseline is then zeroed by passing blanks through the manifold, or the desired span is achieved by adjusting "zero" while the low-concentration standard is passing through the system. The circuit could also be used with inverse colorimetry. The cost was about $200 in parts.

Later, Walker and Amador (882) reported an improved version of this circuit, which converted the Colorimeter signal from transmittance to absorbance. The schematic includes an FET operational amplifier, a logarithmic operational amplifier, and several stabilizing circuits. It could drive a conventional Technicon Recorder or a standard two-input laboratory recorder. The circuits suppress noise picked up from the Sampler switches, provide adjustable gain and baseline control, filter 60 Hz noise, and supply the necessary reference voltage (about 25 mV) for the Technicon Recorder. Signals up to 1.2 absorbance units are provided, limited by the selenium cells in the Colorimeter. Readout directly in concentration units may be achieved by adjusting the gain, and the system operates in either direct or inverse colorimetry modes. These authors (882) give complete setup and operating instructions, and state that modular printed circuit board patterns may be obtained from them.

An analog circuit to "regenerate" curves from the Technicon Colorimeter is described by Walker et al. (886). This system not only provides logarithmic conversion and optional scale expansion from the Colorimeter percent transmittance output, but in addition it electrically counteracts (corrects for) sample interaction and reduces the apparent carryover practically to zero. The circuit has been discussed in detail by Walker et al. (883-885) and the theory behind it is described in Chapter 7.

6. Flowcells

Hamilton (372) presents a useful discussion of flowcell design, with references to the literature. Werner (897) shows a photograph of a Technicon 15-mm Colorimeter flowcell holder, enlarged to accept a 20-mm flowcell. Goodall and Davies (335) made a 10-mm flowcell of 1.2-ml internal volume for the Colorimeter. The flowcell was made from a piece of flattened tubing and did not require plane polished windows. Browett and Moss (139, 140, 610) used a special Colorimeter flowcell in which the liquid flowed smoothly upward through the optical area. This design overcame a problem encountered with the standard flowcell, that of film formation on the end windows.

As mentioned earlier, mixing in the Colorimeter debubbler is a major source of sample interaction or carryover. To gather data on the magnitude of this effect, Habig et al. (356, 357) designed a special "bubble-gating" flowcell for use in the Colorimeter. The flowcell was equipped with electrodes at its entrance and exit to permit detection of air bubbles in the flowcell via the large decrease in conductance when one or both electrodes came in contact with an air bubble, and the body of the cell was made of Teflon to accentuate changes in conductivity. The electrodes actuated a relay (the circuit diagram is given in their article) which removed power from the recorder's servo motor, and this action effectively suppressed noise spikes on the recording whenever bubbles were present. A Heath EUW-20A recorder was used to monitor changes in absorbance too rapid for the standard Technicon Recorder to follow. Habig et al. demonstrate that sample interaction decreases as more of the liquid entering the debubbler is pulled through the flowcell and that sample interaction is minimized when the entire stream including bubbles is sent through the cell. By doing so, they were able to operate a manifold, normally run at 50 samples per hour with 5% interaction, at 180 samples per hour with 4% interaction. Another advantage of this system lies in the preservation of the stream bubble pattern, permitting the addition of more reagents for further reactions downstream of the Colorimeter, or permitting the reading of the same stream's absorbance again after passing through a delay coil for kinetic measurements. Habig et al. (356, 357) did not recommend routine use of the bubble-gating flowcell system for several reasons: Variations in the Proportioning Pump roller action yielded varying amounts of sample or reagents in adjacent segments, which led to noisy steady states; the Sampler timing was not accurate enough to fully exploit the system; and the flowcell would require further design refinement to make it rugged enough to function reliably under long-term use and abuse.

Gochman (331) reported a system for counting blood cells, which incorporated a special flowcell, a reverse-flow wash valve, and a timer. The valve was used to backflush the cell for 10 sec after each sample to prevent accumulation of blood cell debris. Saunders et al. (720, 721) describe a special "sheathed stream" flowcell used for reading blood cells suspended in the flow stream. The sample is fed in as a narrow stream, and a "sheath" solution is added through a special fitting, flowing around the sample stream and enclosing it cylindrically. The sheath holds the cells in focus optically but does not confine them. This design reduced flowcell clogging. A specially designed colorimeter was used to record both the absorbance measurement and the cell light-scattering characteristics.

The use of Colorimeter "triple-pass" or "multipass" flowcells, with effective pathlengths of 45 to 47 mm, is mentioned in several papers (222, 225, 300, 876), usually without specific description of the cell itself. Gaddy (295) made a triple-pass, 30-mm flowcell by attaching wedge-shaped mirrors to a 10-mm flowcell.

7. Debubblers

Irvine (426) shows a debubbler made from standard Technicon fittings and tubing for use in the Colorimeter. It was designed to obtain maximum flow to the flowcell, while reducing the risk of air bubbles accidentally reaching the flowcell. Abdullah and Riley (2) also used a special debubbler in the Colorimeter which improved the shape and reproducibility of the peaks and which permitted fast flow rates through the flowcell. The debubbler is a 0.5-ml glass sphere, with three entrance and exit tubes placed in a single plane. The segmented stream enters the fitting through a horizontal tube, and the debubbled solution flows vertically downward to the flowcell through the second tube, which is at a 90-deg angle to the entrance tube. The overflow leaves through the third tube, which is inclined upward at a 135-deg angle to the other two tubes. *

---

*Abdullah and Riley (2) report that the entrance and drawoff tubes are positioned at a 90-deg angle and that the overflow tube is positioned at a 125-deg angle to the other two tubes. These angles pose a geometrical dilemma. I have assumed that 135 deg was intended.

Clements and Marten (187) used expanded scale colorimetry to achieve high accuracy and precision. The recorded noise level also became expanded, reducing the precision of peak readings. In order to study the effect of the length of the segments sent to the flowcell upon the noise (precision) of the colorimetric measurement, they removed the debubbler from the Colorimeter. The percent transmittance of each segment was thus recorded along with spike signals from the air bubbles. Consecutive liquid segments of less than 5 sec duration yielded poor reproducibility in percent transmittance measurement, but segments lasting for 20 sec gave percent transmittance values reproducible within $\pm 0.1\%$. Unfortunately, the longer segments gave poorer wash. To achieve a satisfactory compromise between precision (low noise) and wash, they used a series of T fittings to remove liquid and air bubbles just ahead of the flowcell in such a way as to merge segments, thus reducing noise without allowing "unlimited diffusion" which would degrade the wash. In the final manifold design, about 10 segments from the original segmented stream were merged to give one long segment. The series of long segments then passed through a small mixing coil and subsequently through the final debubbler to the flowcell. Relative standard deviations of less than 0.1% were achieved via this technique.

Hadley (359) used an AutoAnalyzer system to extract a colored species from an aqueous phase into a chloroform phase; the chloroform phase was then passed through a Colorimeter flowcell. A figure of a special debubbler fitting appears in Hadley's paper. The fitting is designed to replace the usual C5 debubbler in the Colorimeter flowcell compartment and performs two functions on the flow stream: The air segments are removed, and the chloroform layer is separated and sent to the flowcell while the aqueous layer flows to waste.

8. Interference Filters Used with High-Intensity Sources

Before leaving the subject of the Colorimeter, we mention an interesting article by Heerspink and Op de Weegh (381), who studied the properties of interference filters exposed to various amounts of source lamp radiation. They found that a 442-nm filter showed an ordinary transmittance spectrum when measured on a Beckman DU; but, when subjected to high incident energy from a high-pressure mercury lamp in a recording spectrophotometer, the filter's absorbance at constant wavelength (442 nm) oscillated between about 0.2 and 0.3. The period of oscillation slowed progressively, reaching peaks at about 1, 1.5, 2.5, and 4 min after the start of observation. When a cuvette filled with distilled water was placed between the lamp and the filter, no oscillation occurred; that is, the filter maintained a constant transmittance. Of 55 filters examined, 12 showed these oscillations. We thus conclude that interference filters must be used with care when high-intensity sources are employed.

III. DIALYZER

A. Descriptions and Notes on Operation

White et al. (674) give a description of the Dialyzer, along with maintenance and operational hints. The patent notices (268, 766) may also be
consulted. Holl and Walton (408) show a simplified diagram of the stream
flows through the Dialyzer module, and develop two useful concepts: The
rise and fall of concentration (from sample to sample) in the recipient
stream are less rapid than in the donor stream, and the points of maximum
concentration in the recipient stream lag the corresponding peaks in the
donor stream.

Shaw and Fortune (760) discuss some of the ways of optimizing the
Dialyzer parameters, including consideration of the flow rates in the opposing
streams. Lundgren (554) recommended that the flow rates on both sides of
the membrane be no higher than required for adequate separation of samples,
and found that two standard Dialyzers in series gave better sensitivity and
wash than one Dialyzer equipped with large-bore grooving. Lundgren (554)
also mentions that, in phosphate analysis, "preloading" of the Dialyzer may
be necessary to overcome any absorptive capacity of the membrane, and
cautions against situations leading to precipitation of reagents or sample
constituents, because the resulting membrane fouling may yield erroneous
results.

Holl and Walton (408) point out that dialysis is more efficient for low
concentrations than for high concentrations. Thus, response from a system
using continuous flow dialysis is a curvilinear function of concentration.
Lundgren (554) mentions that the range of concentrations through which
dialysis is nearly a linear function of concentration often may be broadened
by maintaining a high electrolyte concentration in the Dialyzer. In addition,
Lundgren (554) feels that the flow rates on opposing sides of the membrane
should not be matched too closely. A close, but slightly imperfect match
may result in a slowly alternating "in-phase" and "out-of-phase" opposition
of liquid segments, which could lead to periodic variations in the concentration of the dialyzable species in the recipient stream output.

Shaw and Fortune (760) describe a problem which may arise in Auto-
Analyzer systems using the original Sampler (without wash box) with the
Dialyzer. They show a figure of two steady-state recordings. In the first,
the Sampler was stopped in the sampling position long enough to develop the
full steady-state output on the Recorder; then the Sampler was cycled so as
to aspirate air into the manifold. Near the end of the steady-state recording,
the signal increases suddenly, reaches a new "steady state," and eventually
returns to the baseline. The abrupt increase in recorded signal was due to

the compressibility of the "extra air," which slowed down the flow rate in the Dialyzer, causing the relative amount dialyzed to increase. In the second steady-state run, after the sample was aspirated, the Sampler was allowed to cycle fully, stopping on a wash cup. The resulting recording was free of the abrupt change near the end. Shaw and Fortune conclude that, when the Sampler is operated normally, any extra air getting into the system, for example, by accidentally stopping the Sampler in midcycle or by not providing a wash cup at the end of the run, may cause errors in the last few results.

Brown et al. (143) used the AutoAnalyzer for continuous in vivo monitoring of glucose and reported that air bubble segmentation in the Dialyzer led to irregular recordings, whereas smooth recordings resulted when the air was added after the Dialyzer.

Levy and Konig-Levy (537) discuss the problems encountered when feeding a solution from a Dialyzer I (203 cm) into a Colorimeter II. The liquid to air ratios become critical. Their recommended ratios are given in Sec. II, C above.

B. Variables Affecting Dialysis Efficiency

The efficiency of the Dialyzer is a function of three variables: the area across which dialysis may occur, temperature, and the flow rates of donor and recipient streams (474, 475, 674). Clarke and Nicklas (186) point out that, as an alternative to using two Dialyzers in series to increase recovery, one may cut the flow rate in half to double the dialysis time. Baum (66) studied the effect of temperature on the recovery of sucrose in the Dialyzer and found that the recovery increased with temperature, leveling off at equilibrium, when 50% of the sucrose had diffused to the recipient stream. Grady and Lamar (340) similarly studied glucose and, again, the percentage of glucose dialyzed increased with temperature. The change in dialysis rate with temperature was great enough to make temperature control of the Dialyzer very desirable. Scobell et al. (752) also used the Dialyzer in an automated method for the analysis of glucose and reported that the analytical result varied 0.26% with a change of 0.1°C in the Dialyzer module. The Dialyzer heater cycled about every 1 to 3 min at 30°C, and this temperature cycling was a major source of error in the automated system.

The advantages and disadvantages of concurrent versus countercurrent flows of donor and recipient streams through the Dialyzer are discussed by several authors. In his original discussion of the AutoAnalyzer, Skeggs (767) states that when the donor and recipient streams flow through the Dialyzer in the same direction (concurrent flow), less material generally is recovered than when opposing (countercurrent) flows are used. However, the rise time, or the time to equilibrate, is faster in the concurrent arrangement, and this is why it was chosen for use. Lundgren (554) confirmed this conclusion, recommending against using countercurrent flows in the Dialyzer, which lead to bad cross-contamination between samples, even at very slow sampling rates (four per hour). However, Brown et al. (147) successfully operated a small dialysis unit (1.1 ml internal volume per channel) in the countercurrent mode, as part of a continuous (steady-state) system. In some cases, countercurrent flow may offer greater sensitivity by allowing higher rates of recovery across the Dialyzer membrane. Increased sensitivity from countercurrent operation in continuous flow dialysis was reported by Kadish and Hall (450), who used a peristaltic pump and a dialyzer system (without air segmentation) similar to an AutoAnalyzer arrangement. The effect of countercurrent versus concurrent flow in an AutoAnalyzer, operated at 60 samples per hour, is shown in a photograph of the resulting Recorder output by White et al. (674). Poorer wash between samples was observed in the countercurrent mode.

Agren and Garrett (8) develop the mathematics of flow dialysis. Although automated analysis is mentioned, the equations derived are quite general and may be applied to several very useful tasks. One may calculate the relative concentrations (of the dialyzable species) in the donor and recipient streams as functions of flow rates and "cell lengths" for concurrent and countercurrent flow dialysis, and this enables one to determine how long the pathlength must be to achieve a desired recovery in the recipient stream. Their equations show that, in some cases, concurrent flow will recover more of the dialyzable species from the donor stream but that, in others, the countercurrent arrangement will be more efficient, since, in theory, one may recover all of the species in the countercurrent recipient stream if a cell of sufficient length is used and if the flow rate ratio (donor/recipient) is less than or equal to 1. They discuss the use of the equations in deciding which alternative would increase recovery of the dialyzable species in concurrent or countercurrent systems: passing the donor and recipient streams into a second Dialyzer; or passing the donor stream into a second Dialyzer, but using pure solvent as the recipient in the second unit. A simulation of the equations is described in which an analog computer (an operational amplifier circuit) is used to solve for the desired parameters. A difficulty here is that one must know the values of certain constants in the equations, e.g., the diffusitivity coefficient for the dialyzable species and "a constant defined by the properties of the membrane." No experimental techniques to obtain these values are given.

C. Interferences and the Donnan Effect

The concentrations of substances other than the analyte often have a marked effect on the rate of dialysis. Baum (66) analyzed for sucrose in boiler water containing dissolved salts and balanced the osmotic pressure in the Dialyzer by adding sodium chloride to the recipient stream. This enhanced the dialysis of the sucrose. Davies and Taylor (218) analyzed water for phosphate and found that the addition of 4% sodium chloride to the samples increased the dialysis of phosphate fourfold. Fusari et al. (294) employed the Dialyzer in analyzing many different drugs in tablet or capsule form and generally used a placebo (a mixture of all ingredients except the drug) to compensate for osmotic pressure differences.

A better understanding of this problem can be attained by considering the "Donnan equilibrium effect." Babson and Kleinman (51) give a clear explanation of this effect, specifically oriented to the AutoAnalyzer Dialyzer, to account for unexpectedly high results in an automated analysis for iron in serum samples. At equilibrium, the product of the concentrations of ferrous ions and chloride ions is equal on both sides of the membrane. The concentration of chloride ions on the donor side is higher to offset the presence of positively charged protein from the serum. Therefore, the concentration of ferrous ions is lower on the protein (donor) side and higher on the recipient side, which raises the analytical results when protein is present. Babson and Kleinman resolved the problem by adding a neutral salt to both samples and standards. This lowered the relative concentration of protein compared to that of all positive ions and reduced the effect of the protein to a tolerable level. Babson and Kleinman (51) also give experiments which may be used to detect and evaluate the influence of the Donnan effect on analytical results.

In similar experiments, Lott and Herman (552) found that serum protein raised the dialysis rates, and thus the analytical results, of calcium and magnesium. They corrected the situation by increasing the concentration of hydrochloric acid in the recipient stream. This increased the sensitivity of the system and decreased the enhancing effect of protein in the donor stream. Bethune et al. (82) used the System II Dialyzer for following enzymatic reactions which yielded ammonium ions. The ammonium ions were dialyzed and determined colorimetrically. When water was used as a diluent for ammonium sulfate standards or as the recipient stream in the Dialyzer II, variable dialysis rates were found for the ammonium ions. This problem was overcome by the use of 0.9% sodium chloride as diluent for the standards and as the recipient stream in the Dialyzer II.

D. Preparation and Characteristics of Membranes

White et al. (674) give details on the preparation and insertion of Dialyzer membranes. Zajac (930, 931) shows pictures of a frame used to hold membranes wrinkle and puncture free while soaking. The frame also facilitates loading the membrane into the Dialyzer. Zajac has also designed a Plexiglas tank for the soaking operation (930).

Shaw and Fortune (760) state that membranes break down rapidly if acid or alkali is used in the flow streams. This becomes especially acute when acid and alkali are used on opposite sides of the membrane. They recommend addition of such reagents to the postdialysis stream, if possible.

Ambrose (25) reported that certain batches of membranes caused quenching effects in an AutoAnalyzer fluorometric determination of phenylalanine. This problem was solved by soaking new membranes overnight in 0.07% Brij-35 solution and then pumping 0.07% Brij-35 solution through the Dialyzer for 3 hr before starting the analyses.

Payne et al. (650) mention that sample interaction may be affected by differences between membranes used in the Dialyzer.

E. System II Dialyzer

The Dialyzer II is described by White et al. (674) and Smythe et al. (781). The unit is transparent, allowing inspection of bubble patterns throughout the flow paths.

Kenny and Cheng (468) found, in a determination of chloride ion, that greater sensitivity was obtained from a 12-in. Dialyzer II when the donor stream entered the upper chamber and the recipient stream entered the lower chamber than vice versa.

Neeley (621) used a 6-in. Dialyzer II and recommended that donor and recipient streams should flow through the unit at about the same speeds, to get optimum peak heights and sample separation.

Chasson (178) reported that temperature was hard to control on an Auto-Analyzer II Dialyzer module. The temperature of the unit was balanced between heat provided by the internal heating block and loss by the unit to the environment. As the instrument warmed up, response to standards became higher and higher. Better results were obtained by operating the module with the cover off, depending on constant room temperature to stabilize it.

The Dialyzer II has been used successfully with "AutoAnalyzer I" components (24). Kenny and Cheng (468) used 6- and 12-in. pathlength units from an SMA 12/60 with the AutoAnalyzer I to determine carbon dioxide concentration. The substitution of the Dialyzer II for the carbon dioxide trap separator fitting permitted the use of silicone-rubber membranes for direct dialysis of carbon dioxide, and the bubble patterns were always visible for monitoring.

F. Special Applications

1. Equipment Modifications

Several authors have found that the Dialyzer makes a convenient extra "heating bath." The dialysis unit may be removed, allowing substitution of delay coils of various length (256) or two 40-ft delay coils (730). Mixing coils may also be used without removing the dialysis unit (534, 648), and 20- or 40-ft delay coils have been used in addition to two dialysis units in the Dialyzer (250). Capps et al. (166) show pictures and drawings of an assembly to hold a single or double mixing coil inside the Dialyzer bath, along with the dialysis unit. Haney et al. (374) show a rack which supports a 100-ft incubation coil of polyethylene tubing, along with mixer coils, inside the Dialyzer bath.

Annino (33, 34) gives construction details and a drawing for a transparent dialyzer bath for the AutoAnalyzer dialysis unit. It is made from Plexiglas and allows observation of the dialysis unit in operation, including bubble patterns. The increased visibility helps to monitor the condition of the membrane and to trace difficulties to their source.

Jansen et al. (435) constructed continuous flow dialyzers from Teflon. This material gave smoother surfaces and reduced blood sample clot hangups in the grooves. Regular cellophane membranes were used, and they lasted about five times longer in the Teflon dialysis units because clots moved through freely, preventing pressure buildup which would damage the membrane. The article by Jansen et al. contains a photograph of the dialysis units, which were mounted side by side, vertically, for convenience.

2. Membranes

Chemical treatment of membranes may improve their permeability. Clarke and Nicklas (186) give an example of this technique applied to continuous flow dialysis of peptide hydroxamates.

Several special membrane materials have been reported. Seifter et al. (755) used a Teflon membrane, 0.5 mil from DuPont, to dialyze carbon dioxide. Evans and Bomstein (249) describe two "permselective" membranes, one a silicone-polycarbonate copolymer (General Electric MEM-213 film), the other a translucent dimethylsilicone rubber (no identification given). These membranes do not separate substances on the basis of molecular size, as is the case with the standard cellophane membrane, but act on the following mechanism: The substance dissolves in one surface, diffuses through to the opposite surface, and is then available for release into the recipient stream. Thus, solvation in the membrane material is the key step. It appears that substances with high vapor pressure, such as vanillin, methyl salicylate, and most dissolved gases, such as carbon dioxide, hydro-

gen sulfide, or ammonia, are good candidates for nonporous membrane
separations. Sodium ions, potassium ions, and chloride ions did not pass
across the membrane; salicylic acid did, but aspirin did not. The MEM-213
was stronger and better resisted deformation. However, the dimethylsilicone rubber membrane was more fragile and was used with donor and
recipient streams balanced to within 0.1 to 0.2 ml/min. Backpressure
coils were sometimes needed on the donor exit line to balance pressure
caused by the manifold acting on the exiting recipient stream, thus avoiding
deformation of the dimethylsilicone rubber membrane. These precautions
were unnecessary with the MEM-213 membrane. Methods of cutting and
mounting the membranes are given (249). The MEM-213 and dimethylsilicone rubber materials were about 10 and 50 times as expensive as cellophane membranes, respectively.

Seifter et al. (754) give methods for evaluating new materials for use
as dialysis membranes, comparing manual and AutoAnalyzer dialysis techniques. They discuss the effects of membrane thickness and breakage
strength, as well as the rates of dialysis of various classes of ions. Methods
for comparing membranes and for determining dialysis kinetics are also
mentioned.

Stein (791, 792) used the AutoAnalyzer Dialyzer to study quantitative
binding of small species to macromolecules. In these papers, Stein develops
the mathematics needed to obtain the concentrations (free and bound) of the
small species, based on the analytical values obtained under three experimental conditions: (a) The small species alone is dialyzed; (b) the small
species is dialyzed in the presence of the macromolecule; and (c) the small
species is analyzed directly, without dialysis. Other binding values (e.g.,
ΔF^0, ΔH^0, and ΔS^0 of binding) may then be calculated. These articles
should also be very useful when one is considering the theory of interferences in continuous flow dialysis.

3. Pressure Dialysis and Filtration

Summers et al. (815, 816) reported that higher sensitivity may be
achieved by using continuous flow pressure dialysis. The pressure was
created by using reduced-bore exit tubing running back to the Pump, plus
differential pumping rates. For example (816), 0.60 ml/min air plus 1.62
ml/min liquid were pumped into the Dialyzer donor chamber, and 1.20
ml/min total was pumped out. The recipient stream was pumped in the
normal fashion, and a Technicon D-30 membrane was used. Since maintaining constant osmotic pressure in samples and standards was important,
all solutions were made up in 0.9% sodium chloride, and the sequencing of
samples and standards was strictly standardized. By forcing more material
through the membrane, about a fivefold increase in sensitivity was realized,
which allowed smaller sample volumes to be taken.

In a closely related application, the Dialyzer has been used as a filter by substituting a filtration material for the membrane. O'Brien and Fiore (633) forced liquid through the filter by restricting the flow on the "donor" waste line, thus increasing the pressure on the "donor" side. Britt (135) used a Scheicher and Schuell Type A coarse membrane (0.4 μm pore size) in place of the usual Dialyzer membrane. The sample was pumped into the upper plate at 7.8 ml/min, plus 0.6 ml/min of air segmentation, and was pumped out of the upper plate at 0.8 ml/min. The solution was forced through the membrane into the lower plate and exited under its own pressure.

4. Miscellaneous Physicochemical Methods

Holl et al. (407) used the AutoAnalyzer for ultraviolet (uv) spectrophotometric measurements of drugs and contrasted the use of continuous dialysis and continuous filtration in such systems. An arrangement using one manifold to do several various dilution ratios, as needed, is also discussed. Fusari et al. (294) used the Dialyzer to perform "content uniformity" tests (measurement of the drug present in individual tablets or capsules) on many different drug formulations. They note that it is particularly useful in separating drugs from capsule gelatin. Gelatin caused problems in manifolds using extraction coils instead of the Dialyzer, because it formed emulsions or slowly precipitated in the coils when contacted by the immiscible solvent, eventually clogging them. Using the Dialyzer permitted dissolving an entire capsule (or tablet), thus saving sample preparation time. Aqueous solutions were used generally, but for water-insoluble drugs (steroids, etc.) alcohol-water mixtures were employed. Type C membranes were used, ordinarily at $50^{\circ}C$, and placebos (mixtures of all tablet or capsule excipients except the drug) were prepared to compensate for osmotic pressure differences. Precision and accuracy were good, agreeing with the manual methods within 1 to 2% and showing "three standard deviation" ranges of about 1%.

The Dialyzer has been used to aid the study of dissolution rates of solid drug dosage forms (271, 920). By returning the donor stream to the dissolution vessel, the volume in the vessel may be held constant during the dissolution rate measurement, and any solid particles removed from the vessel eventually are returned to it.

Although the normal practice is to dialyze the analyte away from impurities or undesired constituents of the sample, the reverse process may be more practical. Scheuerbrandt (731) dialyzed out unwanted carbon dioxide and carbonates into water free of carbon dioxide. The recipient stream was discarded and the purified donor stream was analyzed. Countercurrent dialysis was used to accomplish this purification.

A good technique for increasing dialysis efficiency is to place a reagent, selected to rapidly complex the dialyzed species, in the recipient stream.

Keeping the concentration of dialyzed species close to zero in the recipient stream helps drive the dialysis process toward completion. An example of this procedure, used in the analysis of serum iron, is given by Kauppinen and Gref (455).

The Dialyzer may be of convenience in gas analysis. For example, Ferrari (262) analyzed for hydrocyanic or hydrochloric acid vapor by mixing the gas, as bubbles, with sodium hydroxide solution and then passing the solution and bubbles through the Dialyzer as the donor stream. Chloride or cyanide ions formed and dialyzed into the recipient stream, which was also a sodium hydroxide solution, but segmented with air bubbles. The dialyzed ions were then determined colorimetrically.

Gitelman (328) used the AutoAnalyzer to analyze for calcium and found that certain levels of ferric ion interfered. To prevent the dialysis of significant amounts of iron, cupferron reagent was added to the predialysis stream, precipitating the iron in a single mixing coil, and the stream was then passed into the donor side of the Dialyzer.

Wrightman and McCadden (922) have described a most interesting combination of continuous flow dialysis plus carefully chosen reagent parameters, which allowed indirect analysis for aluminum. The aluminum was complexed with EDTA and was passed into the Dialyzer as the donor stream. The recipient stream contained a colored reagent, iron-TPTZ [Fe(II)-2,4,6-tripyridyl-s-triazine]. Some aluminum and EDTA passed across the membrane. At the pH chosen, the EDTA preferentially chelated the aluminum. Thus, the dialyzed EDTA chelated all the dialyzed aluminum, and the excess dialyzed EDTA then chelated some of the iron; this reduced the amount of colored iron-TPTZ leaving the Dialyzer recipient chamber. An increase in aluminum decreased the amount of EDTA left over to complex the iron, which resulted in an increased concentration of colored iron-TPTZ and an increased response from the Colorimeter.

IV. DIGESTOR

A. Description and Notes on Operation

White et al. (674) give a description of the Continuous Digestor, along with comments on operation and maintenance. They include tables of revolutions per minute versus drive wheel and motor gear arrangements for 3- and 10-rpm motors, and a table giving approximate conversions of Digestor amperage setting to the resulting temperatures at the first and second stages of the module. Applications described in patent notices (265, 267) may also be of interest.

Several studies of the Digestor parameters have been published (Table 1). These studies generally include the relationships among temperature, helix rotation speed, and reagent strengths, and the effect of these three variables on the "recovery" of the sought element (i.e., nitrogen or phosphorus).

Allen (17) shows the effect of Digestor temperature settings on the recorded steady state of an AutoAnalyzer "Kjeldahl nitrogen" system.

Ferrari (260) and Marten (577) have written detailed explanations of the Digestor module, including drawings of the helix system. Ferrari et al. (269) described experiments concerning the principles employed in the Digestor. Initially, a rotating flask was tried in an otherwise standard Kjeldahl digestion. Two heaters were used: one above, and the other below, the flask. When only the lower heater was turned on, a typical digestion took 30 to 60 min, but when both were activated, the digestion time dropped to 1 to 2 min. They concluded that the ultrathin film of digestion liquid in contact with the hot surface of the flask promoted the extremely fast digestion, and these conditions were used in the Digestor helix. Marten (577) reported similar experiments in a paper on automated determinations of meat protein.

Ott and Gunther (638) and Bide (91) discuss precautions and general safety aspects when operating the Digestor with perchloric acid.

Warner and Jones (891) used faster helix rotation (changed from 7 to 16 rpm) and conducted the digestion at higher temperatures in the last one-third of the helix; these changes improved sensitivity and decreased carryover from sample to sample. Whitley and Alburn (908, 909) increased both the helix rotation speed and the flow rates of liquids to and from the helix, causing the sample and wash liquid pools to occupy longer sections of the helix. They also increased the helix temperature to reduce the viscosity of the liquid in the helix. Both actions improved the helix's washout characteristics.

In using the original Sampler (without the wash box) to feed samples to a Continuous Digestor, Whitley and Alburn (908, 909) arranged a second sample crook to supply a water wash when the sample crook lifted out of a sample cup in order to maintain a constant liquid volume in the Digestor helix.

Gehrke et al. (320, 321) obtained an improved baseline by raising the AutoAnalyzer manifold to the same level as the helix inlet and outlet.

B. Special Applications

1. Equipment Modifications

Lento and Daugherty (530) provided better control over the Digestor parameters by replacing the single-speed helix drive motor with a variable-speed motor.

TABLE 1

Articles Discussing Continuous Digestor Parameters

Field of interest (year published)	Reference
Kjeldahl nitrogen	
(1960)	Ferrari (260)
(1964)	Catanzaro (170)
(1965)	Ferrari et al. (269)
(1966)	Allen (17)
(1966)	Marten and Catanzaro (578)
(1970)	Davidson et al. (217)
In foods (1971)	Lento and Daugherty (530)
In meat (1966)	Marten (577)
In pharmaceutical products (1965)	Tufekjian et al. (848)
In plant materials (1967)	Warner and Jones (891)
In tobacco (1967)	Todd and Byars (845)
Phosphorus	
In biological materials (1969)	Bide (91)
In organophosphorus insecticides (1968)	Ott and Gunther (638)
In serum phospholipids (1964/1965)	Whitley and Alburn (908, 909)

Killer (480) shows adapters and gives techniques for digesting oil samples on the Digestor.

In adding samples to the inlet side of the helix, Sutton and Duthie (817) found that erratic mixing occurred when the sample and digestion acid were led in as separate streams. The acid became hot and caused spurting upon contacting the sample. They devised a "chute" mixer which allowed the sample stream to drop into a cold acid stream and which then carried the mixture to the helix inlet through a Teflon bung. A figure for this arrangement is given in their paper (817).

Problems and solutions connected with sample withdrawal from the helix exit are reported by several authors. Gehrke et al. (317) provided a drawing and description of a special fitting used for sample withdrawal. It consisted of a Teflon tube (0.7 mm i.d.) fitted through a glass tube (3 mm i.d.), the latter having an Acidflex skirt. The inner (Teflon) tube protruded somewhat and picked up the unsegmented sample stream. The outer tube led excess sample to waste, and it was connected to a second tube toward the very end of the helix, which collected waste solution missed by the first. This fitting eliminated the need for a vacuum pump or other means of "controlled partial" vacuum and allowed the use of a simple water aspirator on the waste lines. In other reports, Gehrke et al. (320, 321) improved reproducibility in retrieving the digested sample from the helix exit by replacing the impinger mixer with a jacketed double mixing coil and an A0 fitting. To improve mixing and to eliminate splashing, the water used to dilute the digested sample at the end of the helix was first sent through a "preheat" tube placed along the backside of the helix. The outlet of the delivery tube was constricted so that the heated dilution water was added as a fine, high-velocity stream.

Haff (361) gives detailed drawings of an alternate device for withdrawing samples at the helix exit. The fitting was made from several short pieces of 0.09-in.-i.d. Acidflex tubing connected to a glass tube extending into the helix so as to collect and aspirate the digested sample as well as the overflow material. McDaniel et al. (585) show a special mixing chamber plus a planetary pump arrangement that facilitated recovery of the digested sample from the Digestor and which eliminated variations in the flowing stream due to density differences between the samples and wash.

Another problem which may occur at the helix exit is the formation of condensation on tubes used to add diluent liquids. The dripping of the condensate into the digested samples causes erratic fluctuations in the concentration of the recovered stream. Gehrke et al. (317, 320, 321) preheated water used for dilution by passing it through glass tubing placed on the metal shield of the Digestor and then sent the warmed liquid to the end of the helix, thus warming the glass delivery tubing and reducing the formation of unwanted condensate. They have also used a Fiberglas sleeve on the delivery tube to lead condensate droplets to waste (317).

Kramme et al. (497) operated the Digestor with the heating sequence reversed; that is, the first stage was hottest (420°C), and the second and third stages were cooler (330°C). This sequence helped stop damage to rubber tubing connected to a water aspirator but aggrevated condensation drip problems encountered with the standard fume aspiration tube supplied by Technicon. They designed a special gas aspiration pipet made from 8-mm-o.d. glass tubing, with a ground glass fitting at one end. The pipet had a drip ring (a ridge around its outer surface) and was provided with three holes spaced along its bottom surface; it was somewhat longer than the standard Technicon fitting. The drip ring delivered the unwanted condensate into the helix waste ring, and the pipet's three openings removed vapors efficiently.

Mandl et al. (563) describe a new digestion apparatus which overcomes temperature variations experienced with the Technicon Digestor helix. They also discuss a vacuum system, used to sweep the samples through the digestion tube in the new apparatus, which effectively removes acid vapors through a series of traps before feeding the gas stream to the vacuum pump.

2. Continuous Evaporation

The Digestor module has been used to evaporate organic extracts. Usually a second immiscible liquid phase is sent through the helix, along with the solution to be evaporated, so as to pick up the desired compound, allowing recovery at the helix exit. Thus, Ahuja et al. (10, 13) extracted phenmetrazine into chloroform from alkaline solutions and sent the chloroform phase to the Digestor helix. A stream of dilute hydrochloric acid accompanied the chloroform through the helix and, after the chloroform had been evaporated, the dilute acid solution of the drug was removed at the end of the helix. Feller et al. (258, 259) used a similar approach for ethopabate; they include very detailed drawings of the flow of reagents through the helix and a good explanation of how to set up the system (259). Conaill and Muir (191, 192) and Muir et al. (612, 613) used the Digestor to evaporate ether extracts of estrogens, simultaneously transferring the estrogens to a color reagent. McNeil et al. (590) used the Digestor at $57^{\circ}C$ to evaporate ether extracts of 17-ketosteroids, picking them up in a color reagent; the helix was operated in a high flow of air to minimize the risk of explosion. Hormann et al. (413) blended soil samples with water-acetonitrile solvent and then sent the extracts through the Digestor to remove the acetonitrile. The remaining aqueous phase was recovered at the helix exit.

Most of these continuous evaporation systems are operated at the rate of 10 samples per hour. To increase sample resolution, several workers use a Sampler rate of 20 per hour, loading a wash cup between each sample or standard cup (258, 259, 590).

3. Continuous Distillation

Several interesting applications of the Digestor to continuous distillation have appeared. Ferrari and Smythe (274) arranged for the temperature to increase as the liquid to be distilled traveled down the helix and, as vaporization occurred, the gaseous fractions were recovered through several openings. Gunther and Ott (353-355) added 50% sulfuric acid to a sample of biphenyl, and the mixture was passed through the Digestor. The biphenyl was steam-distilled, and the biphenyl vapors were collected by suction into a funnel placed near the helix exit. Cyclohexane was then added to the gas stream to dissolve the biphenyl. After the stream passed through a cooling coil, it was led to a special "evacuated separator" fitting (a drawing of this fitting is given in their paper), which recovered the cyclohexane phase and discarded any aqueous phase distilled, all under reduced pressure. Von

Berlepsch (874) used the Digestor to eliminate chloride, which interfered in an automated method for uronic acid. Concentrated sulfuric acid was added to the sample, and the mixture was sent to the Digestor helix, where the chloride was removed from the sample stream by distillation.

4. Continuous Concentration

Mislan and Elchuk (602) used the AutoAnalyzer with an atomic absorption spectrometer to analyze for iron, nickel, and chromium in the parts per billion range. They employed the Digestor to concentrate samples before feeding them to the spectrometer. Without modifications, the Digestor could be made to yield concentration ratios of the order of 10 to 1. With their described modifications, automatic preconcentrations of about 70 to 1 were achieved. If long periods of time were used (about 3 hr for the system to equilibrate), factors of 250- or 300-fold concentration were feasible. The concentrated sample was collected from the end of the helix into a chamber and, at timed intervals, the concentrate was delivered to the spectrometer.

5. Continuous Irradiation

Dowd et al. (231) reported the use of the Digestor for photolytic catalysis in the determination of cyanocobalamine. In a preliminary design, they employed a time-delay coil with a light bulb inside and a reflective surround. In the presence of sulfuric acid, light catalyzed the replacement of the cobalt-complexed cyano group of the cyanocobalamine by a sulfate group. The liberated cyanide was then determined colorimetrically. The light and time-delay coil apparatus worked well for clear solutions but, in turbid or nontranslucent solutions, only the layers near the lamp were affected. Dowd et al. then replaced the light and time-delay coil with the Digestor module. The glass helix was illuminated by a series of flood lights, and the thin film of liquid permitted efficient photolytic catalysis.

V. FILTER

A. Notes on Operation

Holl et al. (407) used the AutoAnalyzer to assay for many different drugs by uv spectrophotometry and contrasted the use of the Continuous Filter module with the Continuous Dialyzer module in such systems. Urbanyi and Stober (853) found that peak height and width were functions of the feed rate of the Filter module and of the distance of the filter block from the paper. Ahuja and Spitzer (10) noted that filter paper of very fine porosity caused aspiration of air into the postfiltration line. The use of Schleicher and Schuell No. 410 paper solved the problem. To allow the use of large volume flows, Beyer and Smith (89, 90) modified the Filter mixing block by enlarging its chamber and replacing the connector fittings. Hopkinson and Lewis (412) used a Perspex cover for the filter paper roll on the Continuous Filter module to prevent contamination.

B. Special Applications

Atkinson (44) used the AutoAnalyzer to analyze fatty alcohols and, to keep the samples from solidifying, placed the entire AutoAnalyzer in a cabinet heated to $40^{\circ}C$. The Filter module was fitted with a "rewind" to collect used filter paper and, by gearing down the paper feed to 1 ft/hr, one roll (200 m) of paper lasted for 1 month.

Giovanniello et al. (326, 327) utilized a very interesting version of the ion exchange technique to remove unwanted iron from serum samples before measuring iron binding capacity. They placed a resin-loaded paper (Reeve Angel IRA-410, Type SB-2) in the Filter module, which continuously presented fresh resin to purify the flowing stream.

Heinicke et al. (382) removed the mixer from the Filter and converted it to a "miniature colloid mill," using glass and Teflon parts. The modified mixer was used to prevent the formation of lumps of casein, which were precipitated by the addition of a reagent on the AutoAnalyzer manifold.

The Continuous Filter has been used as a readout device, notably in blood testing systems (810). Sturgeon et al. (811) show a picture of a special eight-stream Continuous Filter module, similar in nature to the standard module but using a roll of wide filter paper. Instead of a colorimeter monitoring the processed samples, the samples were delivered onto the filter paper by the Filter module, resulting in patterns of spots that were interpreted visually. McGrew et al. (586) describe a somewhat similar application of the Filter in screening serum samples for syphilis. The antigen suspension (a reagent causing flocculation with reactive sera) contained charcoal. After being separated through a T piece, the charcoal-bearing clumps were deposited on the filter paper by the Filter module. Reactive sera were easily distinguished by the patterns on the paper strip. Similar systems have been reported by others (738, 739). Peoples et al. (656) have described a commercial version of this concept, the Technicon AutoTyper, which is used for blood-type testing.

Special continuous filtration apparatuses have been constructed for use in collecting the radioactive species in radioimmunoassays. Pollard and Waldron (666) describe an apparatus which strengthened the filter material (Whatman glass fiber) by applying a stream of rubber solution at each edge just before passing the filter material over the filtration (suction) block. After the filtration process, the filter material was dried and then overlaid with adhesive cellulose tape to further strengthen it. The filtration material was then led through a continuous radioactivity scanner device. Bagshawe et al. (52) built a continuous filtration apparatus which used glass fiber paper plus a support of filter paper. From the supply spools, the material passed under a "presoak" delivery head, received the sample, passed under two wash streams, moved past a hot air blower, and finally passed through a pair of scintillation counters.

VI. FLAME PHOTOMETER

A. Descriptions and Notes on Operation

White et al. (674) give descriptions of the Flame I, Flame II, and Flame III modules, along with comments on operation and maintenance.

Isreeli et al. (429) detail the development of the first Flame Photometer module, including an excellent discussion of the operating variables and their effect on the instrument. They cover in depth how the optics and electronics compensate for flame flicker and changes in gas fuel pressure. Levine and Larrabee (532) have presented a good description of the Flame III module.

Gochman (332) and Nevius and Lanchantin (625) describe the operation of the Flame Photometer with natural gas rather than compressed propane. Miller et al. (600) compare results from sodium and potassium assays obtained when using the two fuels, and Mabry et al. (556, 558) found that natural gas gave more sensitivity than propane in sodium and potassium determinations on small volumes of sample.

Hambleton (369, 370) used the Flame III for potassium determinations and mentioned that compressed air from a cylinder gave reduced noise compared to air from a compressor.

Miller et al. (600) recommend pumping solution to the flame, using gravity for waste removal. By so doing, they were able to obtain good data at a sampling rate of 60 per hour.

Neutral density filters are supplied by Technicon and are used to optically balance the channels on the Flame Photometer (674). Sergy (756) found that the physical positioning of neutral density filters affected the instrument's response and used clear optical glass spacers to eliminate the positioning effect.

B. Special Applications

1. Equipment Modifications

Gehrke et al. (319) replaced the factory fuel (propane) pressure regulator with a two-stage regulator to minimize fuel pressure fluctuations and thereby to minimize random recorder drift. They also reported that the Flame Photometer yielded an upward drift on the recorded output when the ambient temperature exceeded 90°F and solved the problem by placing the instrument in an air-conditioned room. Later, Ussary and Gehrke (855) built a water jacket around the Flame Photometer detector block to maintain constant temperature, and this greatly improved the module's stability and resistance to room temperature fluctuations.

Sensitivity increases may be obtained electrically or optically. Elam and Palmer (240) note that the usual range of the Flame Photometer, with its built-in zero suppression, is 100 to 160 mEq of sodium per liter. They explain how to increase the sensitivity to yield a range of 0.625 to 40 mEq of sodium per liter. Hambleton (369, 371) reported a collaborative study of a potassium determination developed with a Flame III module in which the collaborators used various flame photometers. One collaborator reported difficulty in achieving the required scale expansion using a Flame II module, but solved the problem optically using a neutral density filter.

2. Connections to Standard Laboratory Recorders or Computer

Lacy (512) gives a schematic for connecting the Flame Photometer to a standard 10-mV laboratory recorder. Ussary and Gehrke (855) published a schematic for using a Leeds and Northrup Speedomax G Model S recorder with the Flame Photometer. This recorder featured adjustable zero and adjustable range, so that the lower end of the working curve could be suppressed off scale and the upper portion could be expanded to read full scale. In connecting the Flame Photometer to the Leeds and Northrup recorder, they included a 2500 Ω potentiometer between the sample and reference input leads. By adjusting this potentiometer, the effect of pump pulsations on the recording could be minimized.

Griffiths and Carter (349) interfaced the Flame Photometer to a computer; their solutions to various problems are discussed in Chapter 6.

3. Lithium Determination

Although the Flame Photometer is normally used to analyze for sodium or potassium with lithium added as an internal standard, the wavelength filters may be rearranged to allow analysis for lithium (296, 625).

VII. FLUOROMETER

A. Effects of Surfactants and Temperature

Before we consider the AutoAnalyzer Fluorometer modules, two problem areas of continuous flow fluorometry deserve special attention.

First, many compounds commonly added to reagents to improve the flow dynamics in AutoAnalyzer manifolds may yield high fluorescence signals of their own. The amount of interference, of course, depends on the excitation and emission wavelengths (or the ranges of wavelengths passed by the selected primary and secondary filters). Detergents (132), Brij-35 (27), and wetting agents in general (581) should be used with caution. White et al. (674) recommend against the use of detergents to clean the Fluorometer flowcell. Ambrose (27) mentions the leaching of plasticizers from tubing as a possible contribution to high fluorescence backgrounds.

Second, fluorescence intensity is a function of temperature.* The magnitude is often of the order of 1 to 2% per degree centigrade (137, 394); therefore, it is good practice to provide reliable temperature control in the fluorescence measurement step to improve the precision and accuracy of the analytical system (27). Rush et al. (702) provided better control over temperature fluctuations by insulating the transmission tubing leading the flow stream from a heating bath to the Fluorometer II flowcell.

B. Descriptions and Notes on Operation

White et al. (674) describe the Fluorometer I and the Turner Model 111 fluorometer units.

Rather than adjusting the slits or sensitivity settings on the Fluorometer, Weber et al. (894) used neutral density filters passing 1, 10, or 25% of the fluorescence, as needed, depending on sample concentration. The next-to-highest standard was aspirated, and the filter giving the desired response on the Recorder (about 50% full scale) was selected.

Narrow (selective) ranges of wavelengths may be obtained using special interference filters. Sansur et al. (716) used such a filter, which peaked at 440 ± 20 nm, in the secondary (emission) position; the filter was obtained from Laser Optics, Inc., Danbury, Connecticut.

When the excitation and emission wavelengths are close together, it may be difficult to reduce scattered light reaching the detector to a level

---

*The observation that fluorescence intensity is a function of temperature is made by many authors (see, for example, Refs. 377, 491, 617, 674, 702, 864, and 865).

low enough to distinguish the desired fluorescence signals. In other words, the excitation energy, reflected or scattered from the flowcell into the emission light path, may not be effectively removed by the emission (secondary) filter. The result is a high "blank" which decreases the sensitivity of the system. Strickler et al. (803, 804, 807) described a useful solution to this problem. They placed a pair of polarizing sheets in the Fluorometer, one with the primary filter and the other with the secondary filter. The polarizing sheets were placed in the "crossed" orientation, thus reducing the amount of scattered light reaching the detector.

C. Special Applications

1. Equipment Modifications

Cappellina and Cappelletti (165) modified the Fluorometer by adding a mobile slide to the range selector control slide. This permitted continuous adjustment by diaphramming any of the four apertures.

Substitution of source lamps can achieve highly selective excitation energy. Hainsworth and Hall (364) used a thallium lamp in the Fluorometer, without a primary filter, to provide excitation at 535 nm. Muir and Ryan (611) used a thallium light source in a Locarte Co. Mk V fluorometer in an automated method for estrogens in urine.

2. Flowcells

Bourdillon and Vanderlinde (125) described a special flowcell for the AutoAnalyzer Fluorometer (probably Model I). It was made from soft glass tubing, about 3.5 mm i.d., and was about 50 mm long. It was used with No. 4 slits and was mounted with adhesive tape wrapped around the top and bottom of the tube and positioned so it faced the middle of both slits. The flowcell outlet was led outside the Fluorometer, and the effluent was allowed to drip freely into an open container to avoid backpressure. The flowcell was about one-half the internal diameter (about one-fourth the capacity) of the normal Technicon flowcell, and the smaller size improved its wash and carryover characteristics. Fingerhut et al. (277) also describe an improved flowcell for the Fluorometer I made from glass tubing. It was 7 cm long, 6 mm o.d., 3 mm i.d., and its capacity was 0.35 ml. Kabadi (445, 446) shows a photograph of a small-diameter, Pyrex glass flowcell designed for the Fluorometer I (Turner Model 111). Ambrose et al. (28) furnish photographs and diagrams of a flowcell and door kit for the Fluorometer I (Turner Model 111) which increases the sensitivity by a factor of 10. This kit also improves the temperature stability of the system around the cell.

Hochella (400, 401) designed a silvered cuvette with windows cut in the coating to allow passage of light in and out of the cuvette so as to match the beam configuration of the Fluorometer. The silvered cuvette gave a five-fold increase in sensitivity compared to an unsilvered cuvette.

Rush et al. (702) show a drawing of a flowcell designed for the Fluorometer II, which worked well with the slower flow rates used in the Auto-Analyzer II system. The internal diameter of the flowcell was 2 mm at the excitation site, compared to 4 mm in the usual flowcell.

3. Electrical Modifications and Computer Interface

Rush et al. (702) changed the electronics in the Fluorometer II to give independence between the full-scale and blank controls. This independence of the controls facilitated the use of an auxiliary digital printout device. Alber et al. (14) describe a circuit used to interface the Fluorometer II to a digital computer.

4. Nephelometry

The Fluorometer may be used to monitor a flowing stream whose turbidity is made to increase (or decrease) with the concentration of the sought constituent. The "primary" light is scattered off the turbid stream, and some of it reaches the "secondary" optical train and falls on the detector. Filter combinations reported for such use are as follows: primary, 485 nm, secondary, 495 nm (236); and primary, 470 nm, secondary passing above 465 nm (482).

Killingsworth and Savory (481) used the Fluorometer (probably Model I) as a nephelometer (primary filter, 360 nm; no secondary filter) and chose a square quartz flowcell with three polished sides (Hellma Cells, Inc., No. 176F-QS). This cell provided better optical characteristics for nephelometry than the standard tubular flowcell by decreasing the amount of spurious light-scattering. The background light-scattering with the standard flowcell was so severe that the Fluorometer slits had to be closed to the point where sample peaks could not be detected. The square cell did not show this background scattering. However, the wash characteristics of the square flowcell necessitated the placement of two wash cups between each sample or standard cup at the sampling rate of 50 per hour, 1:1. Thus, the effective sampling rate was about 17 per hour. A 20 per hour, 1:6 sampling cam could also be used. The calibration curves of peak height versus concentration were nearly linear in this system.

VIII. HEATING BATH

A. Descriptions and Notes on Operation

White et al. (674) describe the Heating Bath module and give hints on its operation and maintenance. They recommend white mineral oil, U. S. P. or Risella 17, for use as bath fluids. Arbogast (40) found General Electric

10C transformer oil to be satisfactory as a fluid in the Heating Bath; this oil showed less charring than the usual mineral oil or ethylene glycol fluids. Keay and Menage (459) preferred glycerol to oil as a heating medium because the latter fumed after aging.

The flow of alkaline solutions through the Heating Bath gradually dissolves the glass coil (171, 172). Davis et al. (220) studied this effect after noting that markedly different calibration curves were obtained when changing from one coil to another. This was traced to the difference in residence times between new, unused coils and old, well-used coils. The internal volume of a new coil was found to be 23 ml, compared to the 49-ml volume of an old coil. At $22^{o}C$, the transit times varied from 5 min 22 sec to 11 min 29 sec and, at $95^{o}C$, the transit times varied from 2 min 29 sec to 5 min 28 sec, from new to old coils, respectively. The wall thickness of a used 1.6-mm, 40-ft coil ran from about 1.0 mm on the inlet to 0.3 mm near its center. Coils with larger internal volumes were more prone toward surging or intermittent flow patterns.

In selecting an optimum temperature setting for the Heating Bath, one should remember that, as the temperature is increased, the residence time in the heating coil decreases because of the increased expansion of the segmenting air bubbles (63). An idea of the magnitude involved here may be gained from considering the transit times given above from the paper by Davis et al. (220). Since transit time decreases as temperature increases, it is entirely possible that increasing the Heating Bath temperature beyond a certain value may actually decrease the sensitivity of the system.

Marten (576) and Craig et al. (200) point out a convenient technique for determining whether additional heating time (e.g., an additional Heating Bath coil) will improve or degrade the sensitivity of an AutoAnalyzer system. A sample or standard is aspirated until a full steady-state signal is obtained. The Proportioning Pump is switched off for a short period of time, and then restarted. The resulting change in the steady-state signal, if any, indicates whether additional time in a Heating Bath is desired.

The increased bubble volume allows use of coils in the Heating Bath whose internal diameter is somewhat larger than that of coils used at room temperature. This concept was utilized by Smythe et al. (781) in the design of the SMA 12/60 Heating Bath. There, the coil internal diameter was increased to 2.4 mm, and satisfactory occlusion of the tubing by air bubbles was still maintained. The coil length could then be reduced without sacrificing heating time. Ertingshausen et al. (245, 246) used a 2.2-mm-i.d. Heating Bath coil and improved the stability of the bubbles entering and leaving the coil by tapering it at either end. Ferrari et al. (270) reported difficulty in maintaining an effective bubble pattern through two large-bore coils at $95^{o}C$ and solved the problem by using 2-mm-i.d. Pyrex glass tubing for connections rather than Tygon or polyethylene transmission tubing.

The optimum liquid to air ratios in the Heating Bath module have been studied by various workers. Gardanier and Spooner (306) found that the wash between samples improved as the air bubble rate was increased. Looye (547) used a standard 28-ml coil in series with a 44-ml "amino acid" coil to obtain a residence time of 25 min in the Heating Bath. With this arrangement, 0.4 ml/min of air bubble segmentation was satisfactory, whereas 0.8 ml/min of air led to a noisy steady-state signal. Sawyer and Dixon (724) found that the slowest acceptable flow rate through a Heating Bath coil (95°C) was a total of about 2 ml/min of liquid plus air, with a liquid to air ratio of 4:1. At slower rates, sample separation worsened, and good bubble patterns were difficult to maintain with the expansion and contraction of the bubbles upon entering and leaving the Heating Bath.

As a solution flows through the Heating Bath coil, it becomes partially degassed. The dissolved gases form "extra" bubbles; most of them are assimilated into the air bubbles used for segmentation and thus present no problem. However, when the Heating Bath is used without air segmentation, the appearance of unwanted bubbles at the outlet may disturb the flow dynamics of the remainder of the AutoAnalyzer manifold. Delaney (223) solved this problem by increasing the flow rate through the Heating Bath coil, thus increasing the pressure in the coil enough to hold the dissolved air in solution. Increasing the flow rate was more satisfactory than trying to increase the pressure by inserting fine-bore tubing downstream of the Heating Bath coil (between the Colorimeter flowcell outlet and waste).

Tappel and Beck (821) used a solution of dimethyldichlorosilane to silanize Heating Bath coils. Silanization reduced the accumulation of protein from samples in the coils.

B. Effects of Thermostat Cycling

The cycling action of the Heating Bath thermostat may cause a regular increase and decrease pattern in the response of an AutoAnalyzer system (644). Shaw and Fortune (760) published a clear description of the effect:

> In operation, a segmented liquid stream is pumped into the heating-bath coil continuously at a uniform rate, apart from a momentary break when each roller of the pump leaves the platen. Any expansion or contraction produced by the on-off switching of the heater is superimposed on the general expansion that occurs in the heating-bath coil. This alternate expansion and contraction is apparently sufficient to initiate and maintain a cyclic change in the rate of flow of the liquid stream emerging from the coil. If a diluent or reagent is subsequently added direct from the pump and the total stream is then passed to the colorimeter, the recorded trace will be in the form of a wave . . . . The effect can be minimized, but not eliminated,

by imposing a back-pressure on the stream from the
heating-bath, through the insertion of a short length of
suitable narrow-bore tubing between the heating-bath and
the point at which the . . . reagent is added . . . . Apart
from the temperature effect and the ratio of liquid volumes
per minute at the point of dilution, the amount of cycling
probably depends on such other factors as the ratio of
liquid to air in the stream, its surface tension and the
resistance of the internal surface of the heating-bath coil.
The rapid change in the vapour pressure of water with
temperature at 95°C may provide an explanation of this
cycling effect.

Shaw and Fortune (760) also show a figure of two steady-state traces. The
first, obtained without the narrow-bore tubing, demonstrates the cyclic
wave associated with the action of the Heating Bath thermostat, and the
second, obtained with the backpressure tubing in place, shows the desired
constant steady-state signal.

Ambrose et al. (29) operated an AutoAnalyzer system using two Heating
Baths, one set at 70°C and the other at 85°C. They show figures of the
steady-state signals obtained with four different arrangements. In the first,
the 85°C bath was filled with mineral oil, and the 70°C bath was filled with
water. This gave a cyclic output on the Recorder, with a period of about
100 sec. In the second arrangement, the 85°C bath was filled with silicone
oil and the 70°C bath was filled with water, and here the cyclic output
showed a period of about 170 sec. In the third, both baths were filled with
silicone oil, and the period of the cyclic output was again about 170 sec.
In the fourth and most satisfactory arrangement, both baths were filled
with water; this yielded the smoothest curve with a small cyclic fluctuation.
They concluded that the improvement with water was probably due to the
lower viscosity and greater effective thermal conductivity of water.

In addition to the use of backpressure or different heating fluid media,
three other solutions to the problem of thermostatic cycling in the Heating
Bath have been reported. First, one may insert a variable transformer
into the Heating Bath heater circuit, which then substitutes for the thermostat (317, 400, 401). Control within 0.5°C at 95°C has been achieved by
using such a transformer (289, 617). Second, one may substitute a proportional controller for the Heating Bath thermostat (219, 609, 702, 919).
Finally, one may heat the fluid in a separate laboratory constant temperature bath, pumping the fluid through the Technicon Heating Bath chamber
around the glass coils (31, 32).

Sawyer et al. (727) note that thermostat cycling may affect the timing of
peaks, again due to the alternating expansion and contraction of the bubbles
in the Heating Bath coil.

C. Surging

On increasing the Heating Bath temperature, the tendency toward erratic flow, or "surging," becomes more and more pronounced as the boiling point of the flowing stream is approached. Six solutions to the problem of surging have been reported. The first, and simplest, is to lower the temperature of the Heating Bath as far as possible (727, 869). In the second, the flow rate of the solution exiting from the Heating Bath is brought back under the control of the Proportioning Pump by debubbling and resampling, resegmenting with air, and then adding the next reagent (669). This procedure avoids erratic proportioning of reagent or diluent because of uncontrolled, variable outflow from the Heating Bath coil. Third, one may increase the pressure in the Heating Bath coil (and thus raise the boiling point of the fluid) by restricting the flow at the outlet of the coil. This concept was mentioned above in the quote from Shaw and Fortune (760). Lenard et al. (528) similarly used a Tygon pump tube (0.020 in. i.d.) in the circuit between a boiling water heating bath coil and a Colorimeter flowcell debubbler to prevent surging. Fourth, one may use a device to regulate the pressure at the entrance to the Heating Bath coil. Catravas (172) employed about 6 in. of Tygon tubing (about 0.005 to 0.01 in. i.d.) as a vent to waste on the high-pressure (inlet) side of a Heating Bath coil. The narrow-bore tubing allowed a small portion of the solution being pumped to the Heating Bath coil inlet to flow to waste and effectively regulated the pressure at the inlet. The fifth solution is to select a solvent with a higher boiling point. For example, Schwartz et al. (750) developed a method which, at first, used resorcinol reagent dissolved in alcohol. The alcohol boiled in the Heating Bath. By substituting 50% ethylene glycol as the solvent for the resorcinol reagent, the problem was solved. Finally, one may use a high-boiling, immiscible liquid, in place of air, for segmenting an aqueous stream in the Heating Bath coil. Muir and Ryan (611) used light petrolatum to segment an acidic solution as it passed through a Heating Bath set at 95°C. This technique avoided the surging which occurred when air was used to segment the stream. After cooling the emerging stream, the heavier aqueous layer was recovered through a C1 fitting. Glass or hard polyethylene tubing connections gave better segmentation patterns than did Tygon tubing.

D. Special Applications

Teflon coils may be used in the Heating Bath to avoid the eating away of the standard glass coils when alkaline solutions are pumped at fairly high temperatures (42, 171, 172, 223). However, Teflon coils may be more prone to surging than glass coils (172). Rush et al. (702) preferred Kel-F tubing to glass tubing when using alkaline organic streams. They used a Kel-F coil in a Heating Bath at 50°C for a stream containing isopropanol, potassium hydroxide, and water.

By using two variable-speed peristaltic pumps, one on the inlet side, the other on the outlet side of a Heating Bath coil, Bernhard et al. (77) provided 3.5 atm of pressure inside the coil. This pressure permitted conducting a digestion of an aqueous phosphate solution in the Heating Bath at 125°C.

The Heating Bath has been used to study the effect of temperature on the analytical response of AutoAnalyzer systems. Barrera et al. (63) brought the module's temperature to 20°C and then set its control to 80°C. The increasing temperature was monitored by a thermocouple connected to a separate recorder. Tappel and Beck (821) used the Heating Bath to provide a temperature gradient of about 1°C/min. By balancing a thermocouple against a millivolt source, the temperature was recorded by the second pen of a Technicon Recorder.

Kawerau (457) constructed a small Heating Bath coil by joining two small mixing coils at right angles in the horizontal plane, placing them in the center of the Heating Bath, inside of the standard glass coils. The arrangement allowed simultaneous heating and mixing of solutions and generated less backpressure than the standard coil.

Holub and Galli (409) describe a modification of an SMA 12/60 heating module that proved convenient when scaling down AutoAnalyzer I methods for use with the AutoAnalyzer II or SMA 12/60 systems. They took apart the SMA 12/60 sealed aluminum block heating module, substituted a phasing coil (Technicon 157-0202-04), and placed the system in a metal cabinet. The modified module acted as a small heating bath for a single-channel AutoAnalyzer. They give a diagram of wiring which provides power to the unit from a standard 120-V socket.

Several authors describe alternate apparatuses for use in place of the Heating Bath. Lenard et al. (528) used a boiling water bath to heat solutions flowing through 80- or 110-ft lengths of Teflon tubing (0.091 in. i.d.). Bradshaw and Spanis (127, 128) found that the temperature of the Heating Bath (about 75°C) was not critical in their AutoAnalyzer system and suggested using a 1-liter beaker placed on a laboratory hot plate to heat the coils. Prall (676) made a heating bath from a 4.5-qt stainless steel beaker filled with glycerol, containing a 20-ft coil of 0.125-in.-o.d. stainless steel tubing. Keay and Menage (459) and Keay et al. (460) performed distillations from a 2-ft steel coil in a specially constructed bath operated at 116 or 120°C with glycerol as the heating medium. The steel coil was used because the hot alkaline solution attacked a glass coil. Wahl and Auger (881) show a drawing of an aluminum block heater, using thermostatic control, which replaced the standard Heating Bath module. It was equipped with a 15-ft Teflon coil maintained at 190°C and was used to flash-distill samples containing fluoride. Copper dust was placed around the coil to serve as a heat transfer medium. The temperature stability of the bath was found to be quite satisfactory.

IX. PROPORTIONING PUMP

A. Descriptions and Notes on Operation

White et al. (674) describe the Proportioning Pumps, Models I and II, and comment on their operation and maintenance. Daily cleaning of the Pump rollers and platen with carbon tetrachloride is recommended. They observe that pump tube snaking may result if the Pump I platen becomes too smooth; this may be corrected by cleaning the platen and then roughening it with fine abrasive or emery cloth. Troubleshooting Pumps for worn or malfunctioning parts with a stethoscope is also described.

A device for leak detection on a Proportioning Pump is useful for fail-safe operation (245, 246). White et al. (674) describe such a device. Two parallel bands are attached to the Pump platen. Leaking fluids short the circuit between the bands, which activates the leak detector circuitry. The device has an outlet provision for sounding a remote alarm.

The level of reagents and samples in relation to the level of the Proportioning Pump may exert rather subtle effects on the performance of an AutoAnalyzer system. Bonk and Scharfe (123) analyzed samples of city water and mention that the height of the standard supply had to be within ± 3 in. of that of the sample supply to avoid miscalibration because of different suction heads at the pump line. Rudnik (700) analyzed boiler water, setting the AutoAnalyzer baseline by aspirating deionized water, and found that the baseline setting varied with the height of the bottle containing the deionized water. Demmitt (225) used an AutoAnalyzer to monitor various process streams and selected the different sample streams by means of solenoid valves. To avoid subjecting the Pump sample line to different pressures (and thus to avoid different sampling rates), a device was used to maintain the sample streams at atmospheric pressure. A drawing of this constant pressure overflow device is given in Demmitt's paper (225).

Placing reagent reservoirs below the Pump may lead to partial degassing; locating reagent bottles above the level of the Pump may reduce the tendency for unwanted bubbles to form in the lines (382). These extra bubbles are especially troublesome in manifold designs which do not use air bubble segmentation. Adelman and Pellissier (4) used such a manifold design to monitor a factory process stream continuously and located the reagents above the level of the Pump to reduce formation of air bubbles in the reagent lines. In order to eliminate the occasional extra bubbles in the sample stream, they segmented it with air and then immediately debubbled it, passing the unsegmented stream to the analytical manifold.

When using the liquid Sampler, the heights of samples and standards are kept reasonably constant. However, the height of reagent supplies must still be considered as potential sources of difficulty. Roland et al. (689)

found that the pump rate of a color reagent was extremely critical. To bring this variable under control, they specified the height of the reagent level in the supply bottle above the AutoAnalyzer manifold. A drop of 2 cm in the reagent level from the specified height caused a change in baseline of 0.0066 absorbance unit. For similar reasons, Noble and Campbell (629) placed all reagent bottles and the flowcell waste line reservoir 10 to 15 cm above the level of the manifold.

Galley (300) obtained high sample flow rates by using two pump tubes in parallel. When both were mounted on the same Proportioning Pump, pulsing and poor bubble patterns resulted. This problem was solved by placing one of the sample pump tubes on a second Pump that was set to operate out of phase with the first Pump.

Summers et al. (815) used a Proportioning Pump to pressurize a Dialyzer so as to increase recoveries (see Sec. III, F, 3). To obtain optimum pumping under pressure, six lines maximum were placed on a Pump, and the Pump with the fewest tubes was used for the backpressure lines.

Smythe et al. (781) discuss the rationale behind the "air bar" on the Proportioning Pump III. By adding air bubbles at the same frequency that the Pump roller leaves the platen, perfectly uniform sample segmentation and reagent proportioning can be maintained in each liquid segment. There are two ways to simulate the action of the Pump III air bar on the Pumps I and II. First, one may use positive air pressure on the air line inlet to the Pump. Prescott (678, 679) used 3 psi of air pressure and reported improved bubble patterns and reduction of backpressure in the manifold system. Christopherson (185) used 10 to 20 psi of air pressure and found that the controlled air segmentation thus obtained aided in uniform segmentation and improved mixing in the manifold. Second, one may arrange a large-bore air pump tube to feed air into a second, smaller-bore pump tube which meters the slightly compressed air into the manifold. Habig et al. (356, 357) used a 0.045-in. pump tube to force air into a 0.020-in. pump tube. The action of the Proportioning Pump rollers trapped a portion of compressed air, then released it, resulting in injection of reproducible air bubbles. The Pump I released a bubble every 3 sec, the Pump II every 2 sec. Similarly, Levy and Konig-Levy (537) used a 0.040-in. air pump tube to feed a 0.025-in. tube, and the latter metered air from the Pump I to the manifold. They show recorded peaks obtained with and without this air injection technique, and pulsing noise was much reduced when the pressurized air was used.

B. Characteristics of Pump Tubes

Smythe et al. (781) show diagrams of the action of the Pump rollers on large- and small-bore pump tubes. They demonstrate that a small-bore tube pumps slightly more efficiently than a large-bore tube, since the roller leaving the platen releases the smaller one slightly ahead of the larger one. White et al. (674) state that pump tubes should have the same wall thickness regardless of their internal diameter to obtain reliable proportioning. This may not be true of certain sources of tubing.

Several authors have commented on the variability of actual pump rates from batch to batch of tubing (169, 317, 689, 747, 787, 791, 792). Jordan (444) found that "identical" manifolds gave different responses to the same standards and ascribed these differences to variation in pump tube flow rates.

It is good practice to measure, or calibrate, the flow rates on all critical pump tubes, and three techniques for doing so have appeared in the literature. The first and perhaps most straightforward procedure is to measure the time required for the pump tube(s) to fill a calibrated vessel. The inlet side may be checked by attaching the tube to a 1-ml or 5-ml pipet, then dipping the pipet into the liquid and measuring the time required to fill it (802). The outlet side may be calibrated by measuring the time required to fill a volumetric flask. By so doing, Fasce and Rej (256) found that, on the average, each package of 12 pump tubes contained two sets of three tubes which were within $\pm 0.5\%$ in their rate of delivery. Second, one may calibrate pump tubes by weighing timed effluents (237). Finally, a combination of volumetric and spectrophotometric measurements may be used. Strandjord and Clayson (801) give a good example of this technique, using a Beckman DU adapted for flowcell measurements. They placed all the manifold waste lines into a 50-ml volumetric flask and obtained the time required to fill the flask. This yielded the total manifold flow rate, which was 4.915 ml/min. They prepared a stock solution of potassium dichromate (0.4 g/liter in approximately 0.06 $\underline{N}$ KOH). A manual 1 in 25 dilution of this stock solution measured 0.519 absorbance, so the absorbance of the stock solution was

$$0.519(25) = 12.98$$

This stock solution was aspirated into the sample line of the manifold, and all other liquid lines were placed in water. The output stream of the manifold read 1.022 absorbance. Since

1.022 absorbance (4.915 ml/min)

$$= 12.98 \text{ absorbance (sample line ml/min)}$$

the sample line flow rate was equal to 0.387 ml/min.

We next consider some factors which lead to changes in pumping rates during the operation of a manifold. Certain reagents attack the inner walls of the pump tubes, causing them to swell and decreasing the pump rate of the tube (689). Von Berlepsch (874) used a manifold which required a flow rate of about 2.8 ml/min of a solution of sodium borate and carbazole in concentrated sulfuric acid. A single Acidflex pump tube changed its pump rate within a few hours, but three small-bore tubes (0.92 ml/min each) used in parallel lasted for days. Davidson et al. (217) used three large-bore Acidflex tubes in parallel to obtain a nominal flow of about 9 ml/min of 90% sulfuric acid. They give a graph of pump rate versus time which shows an initial rapid dropoff in flow rate (8 ml/min at 10 min, about 7 ml/min after 1 hr), leveling off into a slow decrease (about 6.7, 6.5, and 6.3 ml/min at 2, 4, and 6 hr, respectively). Overnight, these tubes would recover and the next day the pattern would be repeated. Ellerington and Ferguson (241) noted that chloroform caused an Acidflex pump tube to stretch during the first two days of operation; they cut off the excess length of the pump tube after this break-in period. Pentz (654) made fine adjustments on a critical Acidflex pump tube rate by changing the amount the tube was stretched between the Pump end blocks. Martin and Baird (581) analyzed serum samples, comparing them to aqueous standard solutions, and found that the first aqueous standard following the serum samples was often higher than expected. The first serum sample after an aqueous standard was usually lower than it should have been. They attribute this effect to a difference in pumping rate resulting from the change in viscosity from water to serum, and corrected for it by running two cups of the same solution after a change from serum to aqueous standard, or vice versa, using the second value only for calculation. Smythe et al. (781) state that the delivery rates of small-bore pump tubes do not fall off with time and that small-bore pump tubes pump more reliably and proportion more accurately than large-bore tubes.

Several authors have reported hints on the care and maintenance of pump tubes. Brown et al. (146) used standard pump tubes to pump concentrated hydrochloric acid. By pumping dry with air, rather than rinsing with water, these tubes lasted about 1 week. Lionnel (543) used Acidflex tubing to pump concentrated sulfuric acid and recommended against flushing the pump tubes with water before overnight standby; by pumping dry on air, the tubes could be used 8 hr a day for about 10 days before requiring replacement. To prevent shearing of Acidflex pump tubes, Beyer (84) positioned the polyethylene end blocks so that the tubes were secured by the rounded edges of the tracts, instead of under the projections. Beyer also noted that small Acidflex lines would occasionally not aspirate reagents until primed with a syringe containing the solution to be pumped. Gehrke et al. (317) used light machine oil on Acidflex tubes to prevent abrasion by the Pump rollers. Looye (547) used two Model I Pumps in preference to one Pump II. The latter flattened the Acidflex tubing more rapidly. Kendal (466) recommended rotating Solvaflex pump tubes 90 deg daily. Kenny and Jamieson (467) found that silastic pump tubes may be subject to tearing because of tension against the sharp edges of the end block, and they improved tube performance by taping the tubes at these contact points.

Zajac (931) recommends breaking in new pump tubes before use by pumping water and shows a picture of a glass tube, shaped to fit around the Proportioning Pump, that keeps the water at constant level during the break-in operation.

Wilson (913) conducted a study of the lifetimes of pump tubes under continuous use. Tygon tubes held up for at least 300 hr. Over a 5-week test period, pumping tap water, 0.03-in.-i.d. pump tubes eventually split, but larger-bore tubes still performed satisfactorily. Wilson concluded that, for long-term continuous use, manifolds should be designed to use the larger-bore pump tubes because they have longer useful lifetimes.

Several authors (83, 185, 518, 724) minimized flow paths for unsegmented liquid streams by trimming the pump tubes used for such streams as close to the end blocks as possible (see Chapter 4, Fig. 1). Minimization of unsegmented streams reduces sample interaction.

Roudebush (698) notes that materials which may interfere in uv measurements are extracted from Acidflex tubing by chloroform.

Dawes and Tetlow (222) found Esco silicone rubber pump tubes to be superior to Acidflex pump tubes.

White et al. (674) give a list of reagents along with their compatability or lack thereof with standard (Tygon), Solvaflex, and Acidflex tubing. Larsson and Samuelson (522) found that a sample stream containing 86% ethanol attacked yellow Tygon pump tubes. By adding the first reagent, an aqueous solution, to the sample stream before it entered the pump tube, they were able to extend the pump tube life from about 5 days to about 14 days.

The adsorption-desorption of material between the flowing stream and the inner surface of pump tubes is discussed in Chapter 4, Sec. III, C.

C. Displacement Pumping

Many authors have found that the technique of displacement pumping is preferable to frequent changes of pump tubes when solvents which attack pump tubes must be used. Table 2 gives a list of solvents and reagents which have been handled successfully by displacement pumping.

Taylor and Marsh (824, 825) describe the use of polyethylene bags containing the solvent or reagent, placed inside a large glass container (e.g., a 2-gal bottle). The volume surrounding the bag is filled with water, and the container is sealed. Water is forced into the glass container by the Proportioning Pump, thereby displacing the solvent or reagent out of the bag into tubing leading to the manifold. This type of apparatus is especially useful when the solvent or reagent to be displaced is miscible or reactive with water. Most workers, however, use displacement bottles without the plastic bags. When the reagent or solvent is miscible or reactive with water, alternate displacement liquids may be chosen. James and Townsend (432) displaced a mixture of sulfuric acid/ethanol (4 + 1) with cyclohexane. Gauger and White (311) used paraffin oil to displace a reagent containing phosphoric acid and glacial acetic acid, and describe how to treat the oil to remove antioxidants. Mercury has been used to displace dioxane (657), acetic anhydride in N,N-dimethylacetamide (348), and pyridine in acetone (829). In another technique, an immiscible liquid is layered between the solvent or reagent to be displaced, and the water that is used to displace it. This approach is not without its hazards. Hainsworth and Hall (364) pumped 2% quinol in 66% sulfuric acid by providing a layer of dichloromethane over the reagent and then filling the remaining volume in the displacement bottle with water. If the bottle were accidentally jarred, mixing of the water and acidic reagent would result in heat, which would vaporize the dichloromethane and rapidly expel the acidic reagent. Later, direct pumping of the reagent with Acidflex was used to avoid such accidents.

By pumping water (or another "displacement" liquid) out of a displacement bottle, one may pump liquids from the manifold, e.g., from a flowcell (48, 58, 163, 218, 224, 359, 432, 610, 687).

It is important to eliminate all air from the displacement bottle and its liquid lines to avoid pulsing (169, 445, 446).

Several authors have described special solvent storage reservoirs designed for displacement pumping. Kabadi (445, 446) used a 2-liter round bottom flask fitted with a ground glass closure which contained the inlet and outlet tubes. Gauger and White (311) give a drawing of a fitting for a standard 1-liter glass-stoppered reagent bottle designed to facilitate removal of air from the headspace over the liquid. Gregory and Van Lenten (348) show a figure of a special displacement bottle used with mercury as the displacing liquid. The apparatus includes stopcocks for refilling the mercury reservoir and for removal of air bubbles.

McNeil et al. (590) give useful hints for displacement pumping of ether. To allow air bubbles to escape and to get a sharp water-ether layer interface, they prepared the ether displacement bottles the night before use by placing 700 ml of ether and 300 ml of water in 1-liter containers. Teflon membranes were placed in the ground glass stoppers of the vessels to form a good seal.

At times it is useful to maintain equilibration between the reagent or solvent displaced and the liquid used to displace it. Muir et al. (612, 613) used a displacement bottle to pump ether to the manifold, where the ether was used to extract an aqueous phase. After the extraction, the aqueous

TABLE 2

Displacement Pumping of Reagents and Solvents

Liquid pumped	Reference
Acetic anhydride in N,N-dimethylacetamide	Gregory and Van Lenten (348)
Amyl acetate	Ganis et al. (304)
Butanol	Hardeman et al. (377)
Butyl acetate	Chevron et al. (183)
Carbon tetrachloride	Carter and Nickless (169)
	Ruzicka and Lamm (708)
Chloroform	Conaill and Muir (191, 192)
	Davies and Taylor (218)
	Kabadi (445, 446)
	Muir et al. (612, 613)
	Robertson et al. (687)
	Roudebush (698)
Dioxane	Persson (657)
Dithizone in benzene	Stanton and McDonald (787, 788)
Dithizone in carbon tetrachloride and chloroform	Moss and Browett (610)
Dithizone in chloroform	Bano and Crossland (58)
	Delves and Vinter (224)
	Hadley (359)
	Ruzicka and Lamm (707)
Ether	Conaill and Muir (191, 192)
	McNeil et al. (590)
	Muir et al. (612, 613)

TABLE 2 (continued)

Liquid pumped	Reference
Isooctane	Hardeman et al. (377)
Isopropyl alcohol in benzol	Valentini (857)
Isopropyl ether	Strickler et al. (804)
Methylene dichloride	Child and Caisey (184)
Methyl ethyl ketone	Avanzini et al. (48)
p-Nitrophenol and ethanol in chloroform	Campbell and Gardner (163)
Phosphoric acid and glacial acetic acid	Gauger and White (311)
Pyridine in acetone	Terranova et al. (829)
Quinol in 66% sulfuric acid	Hainsworth and Hall (364)
Sulfuric acid and ethanol (4 + 1)	James and Townsend (432)
Trichloroacetic acid in chloroform	Hainsworth and Hall (364)

phase was recovered through a B0 separator fitting and was returned to a water reservoir. Water used to displace the ether was drawn from this reservoir. Ruzicka and Lamm (707) prepared dithizone in chloroform by equilibrating the chloroform with an aqueous solution and then used the remaining aqueous phase to displace the dithizone-chloroform reagent to the manifold.

Carter and Nickless (169) observed that variation in pump tube pumping rates makes it difficult to achieve exactly matched addition and withdrawal rates. This was particularly critical in their manifold design, which involved an extraction into carbon tetrachloride. They described a system using a single pump tube and two displacement flasks which yielded the desired exact match in addition and withdrawal rate of CCl_4. Flask A was nearly filled with water, having some CCl_4 at the bottom. Flask B was nearly filled with CCl_4, having some water at the top. The single pump tube pumped water from the top layer of flask A into the top layer of flask B. This pressurized flask B, forcing CCl_4 out to the manifold through a tube extending to the bottom layer of flask B. Simultaneously, a decrease in pressure occurred in flask A, causing CCl_4 (containing the extracted material) to flow through a separator and then through a flowcell, eventually ending in the bottom layer of flask A. The flasks were 500 ml in volume, and all air was excluded from them to avoid excessive pulsing.

D. Special Applications

1. Equipment Modifications

Several papers report the use of Proportioning Pumps modified to run at nonstandard speeds. Catanzaro et al. (171) used a quarter-speed Proportioning Pump to prolong tube life and to reduce the rate of reagent consumption. Crerand et al. (203) used a two-speed Pump to operate an SMA 12/micro system. Switched to high speed, the Pump ran at 20 rpm, the same as the standard SMA 12/60 Pump III, and high speed was used for washing the system. Switched to low speed, the Pump ran at 5 rpm, which permitted the chemistries to go further toward completion in the manifolds.

High-speed Pumps have been used for several quite different reasons. Fasce and Rej (256) found that the standard Pump could not force sufficient volumes of fluids through a set of delay coils. Rather than using large-bore tubes (which would sacrifice reproducibility) they doubled the Pump speed. This was done by installing a larger gear (1 in. radius) on the motor chain drive, plus a longer chain. Bourdillon and Vanderlinde (125) modified a Proportioning Pump to run at nearly twice the normal speed by replacing the standard 8-tooth sprocket on the motor shaft with a 15-tooth sprocket. This allowed doubling the rate of sampling. Habig et al. (356, 357) describe the use of a special Bodine motor on a Pump I to drive it at three times normal speed. This Pump was used in conjunction with the air bubble injection technique described above and released one bubble per second into the manifold. Habig and Williamson (358) used a 30-rpm synchronous motor (Bodine B2270-60G) in place of the standard motor to achieve three times the normal speed with a Pump I. They give a drawing of an adapter plate needed for this installation, along with the gears and drives needed to accommodate the roller system. This Pump was used as a part of an on-line diluter system for feeding samples to a flame photometer (Instrumentation Laboratory Model 143). The standard Pump I produced pulsing fluctuations in the IL 143 digital readout. With the triple-speed Pump I, these fluctuations were more frequent but smaller. In addition, a shorter lag time was realized between introduction of sample to the dilution manifold and eventual readout, plus a faster rise time to a steady photometer output.

Variable-speed pumps have been used as a part of a liquid chromatography monitoring system (397). Place and Hardy (663) used a powerstat to vary the speed of a Proportioning Pump; slow-speed operation reduced long-term consumption of reagents. Brummel et al. (148) reported an interesting study of the effect of varying the Proportioning Pump speed upon the characteristics of an AutoAnalyzer. On the manifold studied, the actual flow rates were not a linear function of Pump speed. As the Pump was slowed down, the steady-state absorbance decreased (0.74 absorbance at 60% of full speed, 0.61 absorbance at 30% of full speed). They point out that one may use the times associated with the changes in recorded absorbance at various Pump speeds to deduce where on the manifold these changes in color yield take place.

Ingram (424) converted a Pump I to act as part of a two-channel Auto-Analyzer. A second platen and chain-roller assembly was mounted on a wooden block and was driven from an extension shaft on the Pump I. The Pump I thus drove the additional assembly side by side with its own.

Fleet et al. (284) controlled the flow of various streams through a polarographic flowcell by means of a cam assembly which opened or closed inlet tubes via pushrods. The cam shaft was driven by a gear train connected to the Proportioning Pump motor.

2. Alternative Pumps

Several groups have reported the use of pumps other than those made by Technicon in continuous flow analysis. Bomstein et al. (122) used a Harvard Model 600-1200 peristaltic pump (Harvard Apparatus Co., Dover, Mass.). This pump featured dual peristaltic channels, pumping 180 deg out of phase with each other. A figure is given showing a modification of the pump, which allowed handling the eight pump tubes required in their method. Jansen et al. (435) designed a special 12-channel peristaltic pump (available from Cenco Instrumenten, Mij. N.V. Breda, The Netherlands) which prolonged pump tube life and reduced tube pulsing and creeping. A photograph and description of this pump is given. Kamoun et al. (451) used an Ismatec MP 13 peristaltic pump to pump samples through an ion exchange column and to pump the eluate to a colorimetric manifold. Vestergaarde and Vedso (872) found that the pump tubes available for the Proportioning Pump would not stand up to the organic solvents required in their procedure, and instead used a Beckman Model 746 metering pump with Teflon valves to pump solutions through a Technicon Colorimeter.

X. RECORDER

A. Descriptions and Notes on Operation

White et al. (674) give a description of the Technicon Recorder, along with instructions on its operation and maintenance. They also briefly describe the SMA 12 Recorder.

Rose (694) shows a figure of a drawing board and T-square arrangement for measuring results from AutoAnalyzer curves. The chart paper was pinned to the board in such a position that the T-square rule would touch all the control sample peaks, thus mechanically compensating for drift. The standard peaks appeared on the same paper as the sample and control peaks, facilitating plotting standard curves and reading off the sample values. Thiers et al. (831) describe a "reading frame" device which facilitates reading charts from multichannel AutoAnalyzers.

Gindler (325) describes a procedure for constructing a nomograph to allow rapid readings of results from AutoAnalyzer methods using inverse colorimetry, wherein absorbance decreases linearly as concentration increases.

Torud (846) compared the error and precision associated with the Technicon Chart Reader to those from three computer programs applied to typical AutoAnalyzer curves. Blezard and Fifield (115) used results from over 200 tests to evaluate errors introduced in reading AutoAnalyzer charts. They compared the error in reading flat-topped peaks, in which steady state was achieved, to the error in reading sharp peaks, in which steady state was not reached. The error in judging percent transmittance, expressed as a standard deviation, was lower by about one-half for the sharp peaks.

B. Special Applications

1. Mechanical Modifications

Allen (18) constructed custom-calibrated scales, reading directly in concentration, which were mounted horizontally on the Recorder, just above the chart drive platen. These scales were designed to accommodate the crosspiece of a T square. By sliding the T square across the scale so as to align the edge of its rule with the emerging recorded peaks, the concentrations could be read without removing the chart paper.

Sturgeon and McQuiston (813) used a multiple pen Recorder which automatically lifted the pens after the signals from each sample were recorded. The chart paper was then advanced rapidly, and the pens were dropped on a fresh portion of paper, thus clearly separating the records of each sample.

2. Electrical Modifications

Devices may be remotely controlled by attachments to the Recorder slidewire drive shaft. Kadish (449) gives a schematic of a circuit which controls the action of syringe pumps. The control unit is governed by the voltage received from a retransmitting potentiometer on the AutoAnalyzer

TABLE 3

Instruments Adapted for Use with the Technicon Recorder

Instrument	Reference
Flame Photometer	
Instrumentation Laboratory Model IL-143	Amador et al. (23)
Fluorometers	
Aminco Photomultiplier Microphotometer Model 10-280	Titus et al. (843)
Aminco-Bowman Spectrophotofluorometer	Kusner and Herzig (503)
Electronics Instruments, Ltd., Model 27A	Antonis et al. (39)
Locarte	Merrills (593, 595)
Locarte Model LFM/4	Rokos et al. (688)
Locarte Model Mk 4	Tan and Whitehead (820)
pH Meter	
Beckman Zeromatic	Prall (675)
Scintillation Counters	
	Pollard et al. (667)
	Webber et al. (893)
EKCO Model N664A	Pollard et al. (668)
Spectrophotometers	
Gilford Model 300	White and Gauger (902)
Hitachi Model 101-UV-VIS	Looye (547)
Turner Model 330	Krieg and Hutchinson (500)

Recorder. Vought (876) used an AutoAnalyzer to monitor cyanide in factory wastes flowing into a river. A microswitch activated by a cam on the Recorder slidewire drive shaft closed a cutoff valve on the waste outflow line when the cyanide level exceeded a selected value. Once closed, the cutoff valve had to be opened manually. To avoid false shutdowns from momentary fluctuations in the recorded signal caused by particles in the flowcell, a mercury-ceramic time-delay switch was placed between the Recorder microswitch and the remote waste line valve. If the preset cyanide limit was exceeded for 10 sec, the waste outflow valve was closed.

Hunter et al. (420) give a schematic and description of a circuit designed to cut an SMA 12 Recorder sensitivity by any desired factor, to allow reading peaks which would otherwise be off scale, and thus to avoid rerunning samples. The system allows adjustment of sensitivity cuts on each individual channel. Tammes et al. (819) describe an automatic scale contraction device for the SMA 12/60. The device is connected at the Recorder terminal box. Signals less than 50 mV, which will remain on scale, are sent to the Recorder amplifier unchanged. Signals exceeding 50 mV, which would go off scale, are sent through a "divide by two" circuit, then on to the Recorder, allowing some samples which would otherwise yield off-scale recordings to be retrieved. The signal can be reduced by greater amounts if desired, and the schematic plus calibration instructions are given.

The Technicon Recorder has been adapted to accept signals from several instruments other than those of the AutoAnalyzer system. Table 3 gives a list of such instruments, along with references to the literature. Looye (547) used the Recorder with a Hitachi Model 101-UV-VIS spectrophotometer and simulated the reference signal normally supplied by the Colorimeter by use of a voltage source, adjustable from 0 to 12 mV, derived from a 1.5-V battery.

Sawyer et al. (727) have described an adaptation of a Technicon Auto-Analyzer Recorder which permitted direct, automatic calculations of assay results. Their procedure is outlined in Chapter 6, Sec. I,A.

Further information on the Technicon Recorder may be found above, in Sec. II,A.

XI. SAMPLER

A. Sampler I: Descriptions and Notes on Operation

White et al. (674) describe the Sampler I, with comments on its operation, maintenance, and repair. They mention that the Sampler I has a sample to wash ratio of about 2:1 and give a table of the unit's sample and wash times for the three operating speeds. They comment that the polyethylene tube extending from the Sampler I crook should be adjusted and then taped firmly in place to prevent gradual slipping which would result in slowly varying sample volumes. Haney et al. (374) used metal or red transparent covers to protect light-sensitive samples in the Sampler I tray.

The volume of sample placed in the Sampler I cups may influence analytical values (589, 830). For consistent results, one should fill the cups
of the Sampler I to a constant level (674, 880). Thiers and Oglesby (835)
published a thorough study of the effect of fluid depth in the Sampler I cups
upon the recorded peak heights. Below a critical volume, the peak heights
dropped off rapidly, due to insufficient sample in the cup. However, even
before reaching this critical minimum volume, a decrease was noted in
peak heights as initial sample volume in the cup was decreased. Since the
Sampler I crook mechanism moves the sample probe into and out of the
sample cup at a relatively slow rate, the initial height of liquid in the cup
directly affects the duration of sample aspiration. When precautions are
taken to ensure constant initial sample height in each cup, the reproducibility of the Sampler I may equal or even excel that of the Sampler II (926).
Reid and Wise (684) measured sample volumes into the cups to obtain
constant depth and achieved good precision with the Sampler I under non-
steady-state conditions. In one example, the precision measured 1.16%,
as a coefficient of variation, for a system operated to give peaks reaching
only 46% of steady state.

One may use the Sampler I to automatically preset the liquid in each
cup to a constant height by attaching a second probe to the crook mechanism,
positioning the probe to aspirate liquid from the cup one position ahead of
the sample probe, and adjusting its height to the desired liquid level. Each
cup is filled to the brim and is placed on the Sampler I tray. The extra
probe aspirates a small amount of excess liquid, which is pumped directly
to waste, and then pulls air. After the Sampler I indexes, the preleveled
cups are aspirated to the manifold by the normal sample probe (38, 758).

Mabry et al. (556-558) note that the slow action of the Sampler I crook
leads to unreliable sampling from small cups of 0.5 ml volume.

The Sampler I was originally designed to aspirate air as a "wash"
between samples. Although an air wash works satisfactorily in some Auto-
Analyzer methods, it is usually difficult to maintain optimum flow dynamics
in the manifold when the sample line is alternately drawing liquid and air.
The cyclic change from liquid to air may also upset the operation of certain
AutoAnalyzer modules, such as the Digestor (908, 909). To provide a liquid
wash with the Sampler I, many laboratories use a constant level reservoir,
placed next to the Sampler I, so that a second crook may dip into it.* The
second crook, attached to the crook lift mechanism, is aimed away from
the sample crook and is adjusted to dip into the wash solution reservoir
just as the sample crook lifts out of a cup (Fig. 2). Generally, the sample

---

*For further examples of the use of the double crook, constant level
reservoir system, see Refs. 48, 57, 94, 125, 133, 156, 159, 205, 217,
224, 276, 346, 354, 380, 404, 433, 434, 462, 504, 513, 609, 618, 632,
636, 678, 731, 759, 763, 793, 809, 848, 857, 908, and 909.

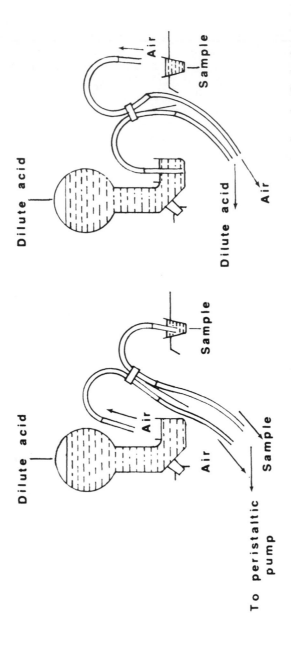

FIG. 2.   Double sampling crook and constant level wash reservoir as used with the Sampler I. In this instance, the wash solution is a dilute acid.   Reproduced from Scholes and Thulbourne (734) by permission of the authors and the Society for Analytical Chemistry.

crook pump line and the wash crook pump line are of the same nominal flow rate, and the two streams are mixed after leaving the Pump. During the sampling cycle, the wash crook aspirates air and, during the wash cycle, the sample crook aspirates air. Thus, a constant flow of liquid and air is maintained throughout the Sampler I cycle. More complete descriptions of this wash technique may be found in papers by Antonis (38), Ferretti and Hoffman (275), and Scholes and Thulbourne (734); photographs of the apparatus are given by Haney et al. (374) and by White et al. (674). Browett and Moss (139, 140) used a constant level reservoir fitted with a small entry port to reduce the risk of accidental contamination of the wash solution. Shaw and Duncombe (758) found that a constant level reservoir did not maintain a satisfactory height of wash solution and gave a diagram of an overflow system, fed by the Pump, which was used instead. Axelsson et al. (49) give a drawing of an alternative device to aspirate a wash solution as the sample tray rotates.

B. Sampler II

1. Descriptions and Notes on Operation

White et al. (674) describe the Sampler II, giving formulas for calculating sample volumes using a 0.030-in. sample pump tube and various cams (10 to 120 samples per hour, sample to wash ratio of 2:1). The patent notice (266) may also be consulted.

MacLean (561) lengthened one side of the Sampler II U-shaped stop pin for ease in handling and painted the pin with white enamel paint and green stripes, to make it more noticeable and less likely to be thrown away with used sample cups.

The faster action of the Sampler II probe mechanism reduces the effect of initial liquid height in the sample cups upon the precision of analytical results. However, for the best precision, some means should be used to ensure fairly constant volumes in the Sampler II cups. Zajac (931) recommends the use of a semiautomatic pipet with disposable plastic tips for this task.

The stainless steel probe commonly used with the Sampler II may cause undesired changes in samples. For example, Stewart (797) found that the stainless steel probe partially deactivated enzyme solutions. Kel-F tubing (18-gauge) was substituted to aspirate the enzyme solutions.

Special segmenting probes are available from Technicon Corporation. These probes contain an extra side arm air inlet fitting, which permits air segmentation at the probe itself. Air segmentation at the probe helps maintain uniform sampling of suspensions (231) and also promotes wash between samples (864, 865). Milbury et al. (598) give a drawing of a segmenting probe (Technicon 127-0247D) used to sample suspended materials in conjunction with the Sampler II stirrer. Bide (91) used a segmenting probe with a large-bore sample inlet to avoid plugging of the probe by particles in the samples. A higher than normal flow of wash through the Sampler II wash box was selected to reduce contamination of the wash solution from sample flowing out of the large probe into the wash box during the wash cycle.

Mabry et al. (556-558) made a "needle probe" for the Sampler II which allowed sampling very small volumes, 150 μl or less from 0.5-ml cups.

To prevent fat droplets from meat samples from forming in the sample line, Gantenbein (36, 305) warmed the wash solution before pumping it to the Sampler II wash box. The wash solution was passed through a double mixing coil, which was wrapped in heating tape and covered with a layer of aluminum foil and a layer of asbestos. The coil temperature was adjusted to about 60°C with a variable transformer.

The Sampler II wash box should be used with caution when organic solvents are present in the wash solution. Antonis (38) reported that the wash reservoir was not resistant to chloroform, and instead used a constant level reservoir flask to hold the chloroform. Dalton and Kowalski (215) found that the Sampler II wash reservoir was resistant to chloroform, but they recommended that the reservoir be washed well with chloroform to remove substances which might interfere with the manifold chemistry. Baird et al. (54) used a stainless steel constant head reservoir to handle a heptane wash with the Sampler II.

To protect the Sampler II cover plate from attack by chloroform solvent in the sample cups, Antonis (38) taped a Teflon disk to the under surface of the plate. Beyer (87) made a sample cover from plate glass for use with solutions in chloroform.

Tappel and Beck (821) published a photograph of a circulating ice water bath designed for use with the Sampler II. Brown and Ebner (145) describe the construction and show a photograph of a water bath made to fit around the Sampler II sample tray. They used a Blue M immersion heater (Model TH-2004) to hold the water bath at the desired temperature. Kline et al. (492) used a circular water jacket around the Sampler II and circulated ice water through a copper coil in an ice bucket, then to the Sampler water jacket, via a Proportioning Pump. Several other papers report the use of similar temperature control systems (68, 427, 520, 661, 666, 814, 822).

2. Imprecision of Cam-Microswitch Timing Mechanism

The Sampler II probe motion from sample cup to wash box is controlled by a cam and microswitch arrangement. The cam drive shaft turns at 10 revolutions per hour. For a sampling rate of 10 samples per hour, the cam makes 1 revolution per sample-wash cycle, and each cup on the sample wheel is sampled by the action of the same cam lobe. Thus, inaccuracy in cutting a 10 sample per hour cam is not serious, because all samples and standards are sampled for the same length of time. Cams designed for faster sampling rates have multiple sampling lobes; inaccuracies in cutting these cams (20 samples per hour and higher) may cause serious reductions in the precision of measurements, because samples and standards will be sampled by the action of different cam lobes. The inaccuracy of Sampler II cams has been noted by several laboratories (24, 345, 356, 357, 897). Such inaccuracies are especially serious when the AutoAnalyzer is being operated under non-steady-state conditions (115, 416, 417, 870).

Several interesting studies of Sampler II cam inaccuracies have appeared. Young et al. (926) found marked variations in timing when different cams of the same nominal value (60 samples per hour, sample to wash ratio of 2:1) were studied in the same Sampler II module. Repetitive variations were traced to the physical variation in shape of the cam edges. When cams of varying sampling times were measured in the same Sampler II module, all the cams showed variations from their nominal rated values, and these variations generally increased with increasing sampling rates. The measurement of cam lobes (20, 40, 60, and 70 samples per hour) were taken and reported, but these measurements did not fully account for the variations in sampling times actually measured. When a single cam was selected and operated in four different Sampler II modules, timing variations occurred from Sampler to Sampler, probably because of slightly different motor speeds.

A key point made above deserves further elaboration: Repetitive variations in timing were traced to the physical variations in the cam edges. One may observe this repetitive variation by aspirating a sufficient number of cups containing the same solution, allowing the cam to run through several revolutions. If a 20 sample per hour cam is used, one looks for up-down patterns in adjacent pairs of peaks; with a 30 sample per hour cam, one looks for repetitive patterns associated with sets of three adjacent peaks, and so on. Often, the valleys between peaks are more revealing than the peak values themselves. If the observed pattern changes when another cam of the same nominal value is substituted, one may be confident that cam inaccuracies are causing the problem. Jordan (444) reports using a correction for cam imperfections based on repetitive sampling of a standard solution through several revolutions of the cam, but Jordan's article gives no details on how the correction was calculated.

Friedman (290) also studied variations in Sampler II cams. In general, the cams showed standard deviations from 0.3 to 0.6 sec. Deviations from nominal sampling times lay between 0.13 and 13.5%, and generally ran between 1 and 6%. The relative standard deviations ranged from 0.5 to 2.5% and tended to increase with increasing sampling rates. As a measure of timing variation from lobe to lobe on a given cam, Friedman used the ratio of the shortest time divided by the longest time, expressed as a percentage. These ratios ran from 98.2 to 80.2% and were generally in the mid-90s. Not much difference was noted between metal and plastic cams. Davidson et al. (217) reported times and corresponding recorded peak values for cams giving 20, 30, and 40 samples per hour, all having a 1:2 sample to wash ratio. Generally, the timing was reproducible within 1% for a given cam lobe (the extreme was 3%), but variations between different cam lobes ranged up to about 6% (the extreme was 10% on a 40 sample per hour cam).

There are several ways to reduce errors from the Sampler II cam system. One may place samples and standards on the Sampler tray in patterns selected to cancel out the inaccuracy of the cam. For example, using a 20 sample per hour cam, Davidson et al. (217) arranged cups in the following sequence: A, B, C, 1a, 2a, 3a, 4a, 4b, 3b, 2b, 1b, C, B, A, where A, B, and C are standards; 1a, 2a, 3a, and 4a are the first of duplicate cups of samples 1, 2, 3, and 4; 4b, 3b, 2b, and 1b are the second of the sample duplicates in reverse order; and C, B, and A are the standards in reverse order. In this pattern, both cam lobes are used on each pair of samples and standards. The average value of each pair was used for the calculations, thus cancelling the effect of cam timing errors. Alternatively, one may use precision cut cams giving timing accuracy to within 0.25 sec (25, 27) or to within 0.5 sec (115, 440). Such cams may be made by carefully filing down the "long" cam lobes (27). Davidson et al. (217) made cams that fit more snugly on the Sampler II cam drive shaft, which further improved reproducibility. Many laboratories disconnect the Sampler II cam microswitch and substitute a high-precision clock timer; this system of timing is discussed in Sec. XI, D, 2, a. Finally, one may select sampling conditions which will allow each peak to more closely approach steady state (306). If each peak reaches steady state, cam timing errors have little or no effect on the analytical result. Theoretical discussions helpful in understanding the effect and magnitude of Sampler II cam timing errors may be found in papers by Evenson et al. (253), Strickler et al. (808), Thiers et al. (832, 833), and Walker et al. (884), as well as in Chapter 7.

C. General Considerations of Sampler Modules

1. Evaporation or Interaction of Samples with the Laboratory Atmosphere

Analytical errors may occur if a liquid sample solution evaporates or interacts with the laboratory atmosphere while standing on the Sampler tray. Two general solutions to this problem have appeared in the literature. First, one may use a liquid layer or a mechanical barrier to isolate the sample solution from the atmosphere and, second, one may provide a controlled atmosphere around the Sampler module itself.

Protective liquid layers were used by Beyer (87), who added a small amount of water after placing chloroform extracts of drugs into the sample cups to suppress evaporation of the chloroform, and by Guillaumot (352), who placed a layer of paraffin oil over serum samples in the sample cups to insulate the samples from the atmosphere prior to an automated pH determination. In his reports of an AutoAnalyzer method for analysis of carbon dioxide in blood, Skeggs suggested covering the samples with mineral oil (770) or a solution of mineral oil in hexane (4 in 5) (768, 774) to reduce gain or loss of carbon dioxide to the atmosphere. Objections have been raised to the use of oil layers in this determination. For example, Gambino (301) published a bibliography of articles concerning mineral oil and carbon dioxide which tended to recommend against the use of oil layers to protect blood samples from gain or loss of carbon dioxide. Others mention that the small amount of oil which is pulled into the sample line at the probe interferes with bubble patterns (365), fouls pump tubes or transmission lines (556, 558), or causes general harm to the AutoAnalyzer system (302, 831). Gambino and Schreiber (302) found that the addition of a drop of 1 $\underline{N}$ ammonium hydroxide solution brought the sample pH to between 8.6 and 8.8 and that loss of carbon dioxide over a 4-hr period was negligible in this pH range.

A mechanical barrier to protect blood serum samples from the atmosphere was reported by Friedner and Philipson (291), who used disks made of acetate-cellulose film, 0.15 mm thick, cut to fit inside the sample cups. By floating on the surface of the serum sample, the disks minimized loss of carbon dioxide to the atmosphere while the samples stood on the Sampler tray. The Sampler probe tilted the disks while sampling. Thiers et al. (831) made similar disks of 0.403 in. diameter from sheets of polyethylene 0.02 in. thick. Mabry et al. (556, 558) found that these disk covers worked well, but others report that the disks do not totally prevent loss of carbon dioxide (302, 831). Other commonly used mechanical barriers to the atmosphere consist of thin metal or plastic films. Afghan et al. (6) placed Saran Wrap around the entire Sampler tray, holding the wrap in place with a rubber band. The Sampler probe was replaced with an 18-gauge needle which punctured the Saran Wrap without tearing it. Thin aluminum foil may be crimped over the cups, allowing the probe to break the seal while sampling (242, 342). Ellis and Hill (242) used a special probe holder fitted with a thumbscrew to hold the probe firmly in position for this purpose. Parafilm has also been used to cover sample cups (342). A sharpened probe is usually required to puncture a Parafilm seal (818). Ryan and Morgenstern (711) used a sharpened 0.034-in.-i.d. probe to puncture Parafilm over 2-ml cups or a sharpened 0.016-in.-i.d. probe to puncture the film over 0.5-ml cups; they placed a 2-cm length of Tygon tubing over the probe at the section clamped by the Sampler II probe arm to prevent the probe from slipping. Taylor and Northmore (823) used Parafilm to cover cups on the Sampler I, and strengthened the crook's descending force by adding an elastic band to the crook mechanism tension spring. A bent hypodermic syringe was fitted to the crook to puncture the Parafilm, and a special bridge was attached to the Sampler I, above the cups, to prevent them from rising out of the sample tray as the crook lifted.   Marten (576) used small sample cups covered with serum caps in a Sampler I equipped with a probe sharpened to a needle point.

The atmosphere over the samples may be controlled by a flow of gas. Cham (176) directed a stream of 5% carbon dioxide in oxygen above the two openings of the Sampler plate to keep this space under the desired conditions. Knowles and Hodgkinson (493) provided an atmosphere free of carbon dioxide around a Sampler II by fitting a Perspex glove box of 120-liter capacity over it.   The box was flushed with carbon-dioxide-free air at 12 liters/min. Samples were degassed and sealed until the analysis was started.   They were then opened and placed in the Sampler tray under the glove box.   One may also use the Sampler itself to degas or equilibrate samples.   Lorch and Gey (551) worked with samples dissolved in heptane.   To sweep carbon dioxide and other gases out of the samples, a stream of nitrogen, previously saturated with heptane by bubbling through a trap vessel, was passed through the samples on the Sampler tray.   To accomplish this, a second crook holding a piece of polyethylene tubing which carried the nitrogen was installed on the Sampler; the tubing was adjusted to dip into the sample cup one position ahead of the sampling crook itself.   Marsters (574) used three additional probes attached to the Sampler crook mechanism; the extra probes equilibrated samples at positions 4, 5, and 6 ahead of the sampled cup with artificial alveolar air.   Zaroda (936) used a similar arrangement, in which two extra probes were positioned to pass artificial alveolar air through the two samples ahead of the cup being sampled.

2.   Errors Caused by Sample Cups and Drift

The sample cups themselves may give rise to analytical errors. Ruzicka and Lamm (707) reported that mercury was lost from solutions by adsorption onto Sampler cups.   Hall and Whitehead (368) found that calcium was absorbed from serum samples by the cups; polystyrene and glass cups were worse in this regard than polypropylene cups.   Auerbach and Bartchy (45) analyzed for quaternary ammonium compounds and recommended rinsing plastic cups with the sample solution to reduce the variation in adsorption of the compounds from cup to cup.

A pattern or grouping of samples and standards may be selected to counteract the effect of long-term AutoAnalyzer drift. If the drift is linear with time, a pattern such as that described by Shaw and Duncombe (759) may be used.   Cups are placed in the order A, S, B, C, D, S, E, F, F, E, S, D, C, B, S, A (or a similar centrosymmetric pattern), where A through F are samples and S represents a cup of the same standard solution.   The averages of each pair of cups, taken about the center point of the pattern, are used in the calculations to eliminate the effects of drift.

3. Further Notes on Operation

It is good practice to minimize the length of line carrying the sample stream from the Sampler module to the Pump (185, 518) to avoid unnecessary mixing which may occur in any such unsegmented stream.

Several authors have commented on the air bubble which appears in the sample line between a sample and the following wash, or vice versa. With the Sampler I, this occurs when using the double crook, constant level reservoir wash system mentioned earlier, because one cannot always adjust the two crooks so as to prevent irregular admission of air as one crook lifts and the other descends. With the Sampler II, the extra air bubble appears during the short time the probe is in air, moving from sample cup to wash box, and vice versa. The extra air from the Sampler line may cause erratic bubble patterns in the manifold and may momentarily affect the recording (389). McCullough et al. (584) found that the extra bubble between sample and wash caused the recorded peaks to be split. The general solution to this problem is to debubble the stream from the Sampler before segmenting the stream with air in the usual manner (133, 139, 140, 156, 389, 584, 725). The length of tubing carrying the unsegmented stream between the debubbler and the point of air segmentation should be kept as short as possible (156).

The extra air admitted into the sample line by the probe motion may cause harmful irregularities in the manifold air bubble pattern. However, during its short residence in the sample line itself, the extra air serves a very useful purpose. If there is a difference in density between the sample and wash solutions, Hinton and Norris (399) point out that "catastrophic mixing" is prevented by the small air bubble separating the solutions. They state that if a compound is present that increases surface tension, leakage past the bubble will increase. Gardanier and Spooner (306) also mention that an air bubble should appear as the Sampler probe moves from cup to wash.

A single Sampler module may be used to feed samples to several manifolds. Thiers et al. (831) show a photograph of a clear plastic stream splitter fitting that divides the input stream from the Sampler into five output streams for a multichannel AutoAnalyzer system. Crowley (206) used two probes, which were adjusted to dip into adjacent cups on the Sampler tray. Each probe fed its respective stream to a separate part of the manifold, allowing two different chemical determinations to be conducted simultaneously.

Santacana (717) gives a figure of a special sampling probe which was useful with samples containing heavy suspended material. The probe incorporates a stream of liquid, supplied by the Pump, to aid in sweeping the sample plus particles up into the sample line. Directly after sampling, a special decantation fitting was used to remove the particulate matter.

In addition to the Sampler water baths mentioned earlier, sample temperature may be held constant by placing the entire Sampler module in controlled environments, such as a chamber kept at 37°C (403) or a cold room (859, 860). Antonis et al. (39) kept aqueous samples and reagents at 4°C during a day's run by encasing a Sampler inside a Perspex box containing a layer of solid carbon dioxide.

In some instances, it is more effective to place reagents in the Sampler cups and to pump the sample continuously to the manifold from a single container. For example, Roodyn (690, 691) placed various substrates in the Sampler cups to study their effect on various enzymes in the continuously pumped sample solution. By using a repeating sequence of ten different substrates in four similar groups around the Sampler tray, Roodyn (692) was able to perform ten different enzyme determinations per milliliter of effluent from a chromatographic column.

Many clinical laboratories conduct blood tests by electrophoresis as well as by the AutoAnalyzer. Carruthers (168) reported a convenient technique to take samples for electrophoresis from serum samples prepared for the AutoAnalyzer. Eight needles were attached to a flexible stainless steel strip. The strip was flexed into a semicircle, and the needles were dipped into eight cups of serum samples in the Sampler tray. Small amounts of the samples adhered to the needles. The strip was straightened and was fitted into the electrophoresis tank so as to apply the samples to the electrophoresis sheet.

D. Special Applications

1. Sampler I

a. Sample Mixers. Naumann et al. (620) show a drawing of a mixing device for resuspending blood cells in sample cups. It is a tube of polyethylene guided by the Sampler I probe, adjusted to blow bubbles through the sample cup just ahead of the cup being sampled.

Nelson and Lamont (624) give a picture of a small stirring motor with a Teflon paddle, attached to the Sampler I crook arm. The stirrer paddle dipped into the cup just ahead of the cup being sampled.

b. Double Sample Trays. Lacy (512) modified the Sampler I to hold a second sample tray accommodating an additional 40 cups. The second tray was mounted over the original tray, and a second sample crook made of 4-mm copper tubing was attached to the Sampler I crook mechanism. Ingram (424) also added a second tray and crook to a Sampler I to permit operation with a two-channel AutoAnalyzer system.

c. Synchronized Reagent Aspiration. It is sometimes desirable to pump a reagent to the manifold only when a sample is being aspirated, rather than continuously. Ferretti and Hoffman (275) used a Sampler I fitted with an extra crook, which was arranged to aspirate a color reagent in synchronization with the samples. The manifold flows were adjusted so that the reagent reached the mixing point (a cactus fitting) just before the sample did; the reagent flowed to the cactus until just after the sample flowed past the fitting. A water wash was supplied by a third crook, which dipped into a constant level reservoir as the sample and reagent crooks lifted, and the wash stream was phased to reach the cactus just after the last of the reagent; this wash further improved the separation between samples. Singer et al. (764) also used an extra crook on the Sampler I to aspirate a reagent only when the samples were being aspirated. The reagent itself was highly colored and, if it had been pumped continuously, lower values of the analyte would have been masked. Their paper (764) shows a flask designed to position a reagent for convenient pickup by the extra crook.

The expense of certain reagents provides further reason for aspirating reagent only as needed rather than continuously. Winter (916) used a Sampler I fitted with two probes and a sample tray containing two concentric rings of cups. The reagent was placed in one ring of cups, and the samples were placed in the other. The double probe aspirated sample and reagent in synchronization, and the two streams were mixed on the manifold. Ott and Gunther (636) used a similar arrangement and described how to achieve satisfactory synchronization of sample and reagent flows on the manifold. Blue and yellow solutions were prepared from water-soluble food dyes; one ring of cups was filled with the blue solution, and the other was filled with the yellow solution. The Sampler I was started and adjustments were made in tubing lengths so as to get the purest, most uniform green color at the point of mixing.

d. Sample Containers and Sample Preparation Techniques. Thiers (830) modified a Sampler I to take serum samples directly from Vacutainers, thus eliminating sample transfer into sample cups. Thiers et al. (831) described in detail a conversion of a Sampler I to handle samples in 12-mm x 100-mm test tubes.

Ott and Gunther (637) used a Sampler I to extract pesticides from material scraped from thin-layer chromatographic plates. The scrapings were placed in sample cups and 2 ml of 30% ethanol was added. The Sampler I mixer paddles were adjusted to mix two adjacent cups, and the sample probe was set two cup positions away from the mixer, so that there was always one cup between the last mixer paddle and the sampling position to allow the scrapings to settle.

e. Modifications Which Increase Sampling Flexibility. Several laboratories have described modifications of the Sampler I which increase its
flexibility. Reid and Wise (684) modified the mechanism to reverse the
action of the sample crook in relation to the drive cam; that is, the time
usually spent in aspirating a sample became the time of air aspiration
between samples, and vice versa. This reversal of sample to wash ratio
resulted in shorter sampling times. For example, at 20 samples per hour,
before modification the sampling time was 120 sec; after modification it
was 42 sec. This sampling time was desired because the sample line was
being pumped at 0.6 ml/min, and only 0.5 ml of sample was available.
Shaw and Duncombe (759) used a Sampler I with a modified cam lever,
which provided a 2.5-min sample aspiration with about 0.5 min of wash.
Thiers et al. (831) modified the Sampler I probe mechanism to give a more
rapid action, thereby reducing variability in sampling, especially at the
faster sampling rates. Kuzel (506) described what is perhaps the ultimate
in modifications of the Sampler I. Mechanical components were added to
provide a large range of sample to wash ratios and a large range of sampling
rates. These changes permitted running analyses at as fast a rate as
possible with low carryover (sample interaction) errors. The sample probe
action was made faster, which reduced the effect of sample height in the
cup on the analytical result. The cost of the conversion was about $200.

2. Sampler II

a. Customized Cam Timing and Electric Timers. Several authors have
reported making special timing cams for the Sampler II. Fasce and Rej
(256) show a drawing of a cam which produces a sampling rate of 10 per
hour, with a 40-sec sampling time and the remainder wash. Hathaway et al.
(379) made a 20 sample per hour, 1:2 cam by removing two opposing lobes
from a 40 sample per hour, 2:1 cam. Laessig et al. (516) give the dimensions for a 40 sample per hour cam, yielding 37 to 38 sec of sampling,
with the remainder wash. Nordschow and Tammes (630) modified a standard 50 sample per hour cam to obtain 20 sec of sampling and 52 sec of
wash. The cam sampling lobes were reduced to a rim length of 0.365 in.
subtending 20 deg. The wash recesses subtended 52 deg. This last example
makes clear how one obtains the desired sampling and wash times when
designing a special cam: Each degree of cam rotation represents 1 sec of
time since, in the standard Sampler II, the cam rotates once (360 deg)
every 6 min (360 sec). Kuzel and Coffey (508, 509) describe a "double
rise split cam" made to allow a variable sample to wash ratio with the
Sampler II.

One may also achieve flexibility in sample timing by substituting slower
or faster motors to drive the Sampler II timing cam shaft. Tappel and Beck
(821) point out that a slower drive motor permits full development of steady-

state conditions in kinetic studies. Young et al. (926) and Evenson et al. (253) described the substitution of a faster timer drive system designed to turn the cam at 1 rpm; this cam speed produced much more reproducible timing at 60 samples per hour than did the standard system.

This brings us back to a problem discussed earlier, the variability associated with the standard Sampler II cam timing mechanism and one means of reducing this variability, the use of an external timer in place of the Sampler II cam and microswitch arrangement. In addition to improved precision, an external timer also permits greater flexibility in selecting sampling rates and sample to wash ratios, thus allowing selection of a tailor-made timing program to suit the needs of the associated AutoAnalyzer manifold. Gray and Owen (345) describe a circuit for the Sampler II which provides for the use of an external timer, replacing the cam and microswitch arrangement. Werner (897) shows a photograph of an electric timer used with the Sampler II to allow varying the sample and wash times independently. Amador and Urban (24) used a Gilford Model 4016 timer, and Strickler et al. (808) used a Flexopulse HG timer to control the probe action of the Sampler II. Digital timers have also been used for this purpose (886). Varley and Baker (870) give the circuit for a solid-state timer used with the Sampler II which provides variable times, in switched increments from 1 to 100 sec in 1-sec steps, for both the sample and wash periods. Kruijswijk and Pelle (501) published a schematic and explanation of a transistorized double time-delay switch circuit. Although the AutoAnalyzer is not mentioned in their paper, the circuit was used to regulate the sampling time and the time between samples with "an autoanalyzing apparatus." Bennet et al. (72) used a computer program and a computer interface (consisting of a driver and relay) to control the Sampler II probe action. The signal times were accurate to 0.02 sec. Taking into consideration the timing errors associated with the interface relay action and those associated with the Sampler II probe arm mechanism, overall timing accuracy was better than 0.5 sec.

b. Other Electrical Modifications. Several authors have installed a switch on the Sampler II which reverses the action of the microswitch controlled by the timing cam (25, 27, 68, 91). For example, with a cam designed to give a sample to wash ratio of 2:1, when the switch is in its "normal" position, the ratio remains unchanged, but when the switch is thrown to its "reverse" position, the Sampler II operates in a 1:2 sample to wash ratio.

Bird and Owen (95) give the schematic of a circuit designed to prevent incomplete or improper cycling of the Sampler II by inadvertent operation of the on-off switch at the wrong moment. The on-off switch is replaced by a four-position switch, allowing operation in four modes: standby, continuous, automatic, and emergency. In the standby mode, the Sampler II probe remains in the wash box. If the switch is changed to standby when a sample is being aspirated, the Sampler II finishes aspirating the sample; then the probe returns to wash and remains there. In the continuous mode, the probe moves immediately to the sample cup and remains there. In the automatic mode, the Sampler II behaves as normal except that it will not start aspirating the first sample until the beginning of the sample cycle is indicated by the timing cam, regardless of the cam's position when the switch was thrown to automatic. In the emergency mode, the Sampler II probe moves to wash immediately and remains there.

c. Sample Mixers. Modifications of the Sampler II mixer have been found useful. Burns (153) controlled the mixer speed by placing a 10 Ω, 2 W potentiometer in series with the stirrer motor. Bjorksten (98) bent the paddles of a vibrating-type mixer so that both paddle blades fit into the same 8-ml cup, and used a variable transformer to control the mixer speed. Eichler (237) describes a mixer comprised of a DC motor and stirrer shaft, which was attached to the Sampler II probe arm.

Neeley et al. (622) reported a totally different approach for mixing samples on the Sampler II tray. Small magnetic stirring bars were placed in each sample cup. The bars were agitated by a second set of magnets attached to a ring fitting around the Sampler II under the sample cups. This ring was agitated by a motor drive, thus keeping all the samples constantly mixed. The Sampler II was fitted with an additional attachment which magnetically retrieved the stirring bars after the samples were aspirated and dropped the bars into a wash receptacle filled with water. This stirring system worked better than the paddle stirrer ordinarily used for mixing samples, because all samples were stirred until sampled, not just the two ahead of the sampling position. Splattering of samples and contamination of subsequent samples by the paddles were also avoided.

d. Sample Containers and Sample Preparation Techniques. Beyer (87) prepared chloroform extracts of drug samples and placed the extracts in shell vials for sampling by the AutoAnalyzer. To hold the shell vials in the Sampler II, two metal disks were attached to the Sampler plate below the plastic sample tray. One disk was provided with 40 holes of 1.3 cm (0.5 in.) diameter to center the shell vials; the other disk was solid and was positioned below the vials to support them.

Ahuja et al. (9) show a picture of a support that fits onto the center hole of the Sampler II tray. The support was designed to hold a series of micro-chromatographic columns. This arrangement was used to purify samples by pouring them through the chromatographic columns; the purified samples eluted directly into the Sampler II cups.

Larson et al. (520) described a novel method for the automatic preparation of diluted samples. Their system used two Sampler II modules. The first Sampler II contained the samples and was operated at 40 samples per hour with a sample to wash ratio of 1:4. The samples were mixed with reagents on the manifold and were then delivered via a stationary probe into empty cups on the second Sampler II, which was operated at 80 samples per hour. Thus, each sample was delivered in equal liquid portions into two adjacent cups on the second Sampler II. About 95% of the original sample reached the first cup of each pair on the second Sampler II, but because of diffusion in the manifold tubing, about 5% carried over into the second cup of each pair. The sampling probe of the second Sampler II was positioned about 30 cup positions away from the stationary probe delivering the samples into the cups, and the two sample dilutions were fed into the analytical manifold from the second Sampler II. The Sampler II modules were synchronized by placing a cup of dye solution on the first Sampler II and switching on the second Sampler II just as the first drop of dye reached it. A wash cup was placed on the first Sampler II to clean out the dye, and the samples were then processed through the system. The diluted dye cups on the second Sampler II were removed before they reached the sampling probe. Larson et al. (520) used this dilution procedure to differentiate between strong and weak positive responses of the samples.

e. Intersample Bubble Detectors. The bubble which forms in the sample line as the Sampler II probe passes from sample to wash, and vice versa, may be monitored either electrically or optically, and such monitoring systems have been put to several ingenious uses. Hansen et al. (376) devised an electrical system to detect the interruption in sample line bubbles caused by clots plugging the sample line. The sample line was replaced in two short sections by short pieces of stainless steel tubing, placed close together and separated by a very short piece of plastic tubing. A 1-kHz tone was fed across the two sections of metal tubing and, at 50 samples per hour, with a sample to wash ratio of 1:1, a bubble formed and passed between the metal tubes once every 0.6 min. As a bubble broke the circuit, the electrical pulse was used to hold a time-delay circuit in its normal mode. If a bubble failed to pass between the metal tubes for more than 1.4 min, the circuit tripped an alarm. The circuit diagram and a full explanation is given by Hansen et al. (376). Melley et al. (591) described a clot detection and warning device which used a small photodiode to monitor the amount of light passing through the tubing leading the sample stream from the Sampler II to an SMA 12/60 system. The photodiode was connected to an amplifier which put out an audible beep whenever a bubble passed by the photodiode. At 60 samples per hour, 120 bubbles per hour normally formed at the Sampler II probe, and a rhythmic pattern of beeps was created. Clots caused a series of very short beeps, as did extra air introduced by an accidental displacement of the sample probe or an insufficient sample volume. The cost of parts for this circuit was about $20, and the schematic is given in their paper (591). Young et al. (926) also used a photodiode to monitor the line from a Sampler II probe. The line was passed between a light source and the photodiode, and the volume of each sample could be determined by measuring the time between air bubbles. Both trailing and leading edges of the air bubbles were detected to eliminate timing errors due to different sized air bubbles. The photodiode was used to trip an electronic timer accurate to hundredths of seconds. This method of timing was used for two purposes. First, the time (volume) of each sample was compared to that of the standards. Care was taken to fill each standard cup fully so that a reliable standard time (volume) base could be established. A deviation from this standard time base measured for a sample indicated an insufficiently filled sample cup, and an alarm was sounded if the sample volumes varied from those of the standards by a predetermined amount. Second, the bubble timer system was used to study the variations in sample timing due to imprecision in the Sampler II cam timer system. An Airborne Instruments Laboratory sampler was similarly studied. The use of a wetting agent was found to reduce or eliminate that portion of sampling time variation caused by changes in viscosity between samples and standards. Jansen et al. (435) used a system of continuous flow analysis in which sampling was based on volume rather than time. Two platinum electrodes were placed in the line from the sampling module. As the probe moved into a sample, an air bubble preceded the sample stream down the sample tube. When the bubble passed the second platinum electrode, the circuit was broken, and the sampling module was triggered to move directly to the next cup. Wash cups were placed between sample cups to reduce sample interaction. A new bubble formed as the probe moved, and the cycle continued. In this way, the samples were taken by volume and no sampler timer as such was used. The sample volume was controlled by the internal diameter and length of the tubing between the sample probe and the second platinum electrode. The use of a volume base for sampling produced better analytical precision than that of time-based sampling. In one determination, the precision, as a coefficient of variation, was 2.0% on a time basis and 1.2% on a volume basis. The sampling circuit was also provided with a time relay, which was set slightly longer than the time required to take a sample by volume. If a clot blocked the sample line, an alarm was sounded. Empty cups were also detected in this fashion.

f. Fraction Collection and Delivery of Reagents to Sample Cups. The Sampler II has been used to collect fractions from a chromatographic column (68) or from the outflowing stream of a fluorometer flowcell (822). Hanok and Kuo (375) used the Sampler II to collect small amounts of samples from an SMA 12 Sampler. The Sampler II then fed the collected samples to a second AutoAnalyzer system for additional analyses. This system of sampling for the second AutoAnalyzer was selected in preference to splitting the sample stream as it left the SMA 12 Sampler, because the SMA 12 Sampler's sample to wash ratio was fixed and its ratio was not optimum for the additional AutoAnalyzer system. Hanok and Kuo (375) describe a modification of the Sampler II stirrer assembly to obtain complete sample collection into each cup, one position ahead of the sampling probe.

In a somewhat related function, the Sampler II may be used to hold sample cups in position under a delivery probe while reagents are automatically added to each sample. Irvine and Marwick (427) added two reagents pumped through fine-bore needles into the sample cups. The time of reaction was 22 min, and this incubation time was determined by the time required for the Sampler II to index the samples from the reagent delivery probes to the sampling probe position. Suba (814) used a Sampler II plus a Proportioning Pump to deliver reagents into sample cups already containing samples. The prepared samples in the Sampler II tray were then transferred to a second Sampler II which fed the samples to the analytical manifold. Vargues et al. (868) modified a Sampler II to hold a second sample tray about 13 cm above the normal tray. A portion of the manifold was used to pump reagents drop by drop into 40 cups placed in the upper tray. The samples were then added to the cups manually, and the tray was set aside to complete the reaction. Later, the sample tray was placed in the lower position, where the normal Sampler II probe aspirated the prereacted samples into the analytical manifold. In operation, one set of prepared samples was analyzed on the lower tray while 40 new cups were being prepared simultaneously on the upper tray.

g. Extractions. Samples may be extracted with an immiscible solvent directly in the sample cups. Lorch (550) used two Sampler II modules for this purpose. The first, operating as a normal Sampler II, fed samples to a special device set next to the second Sampler II. This device was similar in action to a Sampler probe arm mechanism, but was built with enough strength to support two motor-driven stirring rods and a set of delivery tips. The delivery tips were used to add samples from the first Sampler II as well as solvents and reagents to the cups in the second Sampler II. The device moved the stirring rods and delivery tips into place during the processing of samples and back into a wash reservoir between samples. The action proceeded as follows: The extraction solvent (n-heptane) was pumped into a glass cup on the second Sampler II. The sample was delivered into the cup and the mixture was stirred to effect the extraction. The Sampler II indexed, and in the next position a reagent was added to the cup to promote separation of the liquid phases. The Sampler II indexed again, and in this position a second stirrer helped to promote mixing and further phase separation. To allow complete phase separation, 15 additional positions elapsed before the sampling probe aspirated the extract.

Bjorksten (98) also modified the Sampler II to perform extractions. Two additional stationary probes were attached to deliver liquids into the sample cups occupying the fourth and fifth positions ahead of the sampling probe position. The Sampler II vibrating-type paddle mixer was set two positions ahead of the sampling probe. In operation, one of the stationary probes added the solvent (nonane-isopropanol, 4:7) to an empty cup five positions ahead of the sampling probe. The sample tray indexed, and the sample was added to the cup at position 4 through the second stationary probe. The tray again indexed and, at position 3, the mixer was activated to extract the sample. After indexing, no action took place at position 2 so that the phases could separate. After the sample tray indexed again, the upper organic layer was aspirated into the analytical manifold. Bjorksten (98) shows photographs of the Sampler II, which was modified to hold funnels under the stationary probes to catch liquids from the probes when no cup was present. The mixer blades were bent so that both fit into the same 8-ml sample cup, and the mixer power came from a variable transformer to allow adjustment for optimum extraction conditions. The rotating type of mixer was found unsuitable. The samples themselves were fed to the modified Sampler II by a normal Sampler II operating at the same speed and sample to wash ratio as the extractor Sampler II, 30 samples per hour, 1:2. The long wash facilitated synchronizing the two Sampler II modules.

h. Synchronized Reagent Aspiration. One may economize on reagent consumption by aspirating a reagent only as needed with the Sampler II. Generally, a second probe is attached to the Sampler II probe arm, and the probe is adjusted to aspirate the reagent in synchronization with the sample (65, 534). If the same wash solution is suitable for both sample and reagent washing, the arrangement is simplified considerably. Webber et al. (893) placed two conical cups in each macrocup, so that a pair of cups indexed under the two Sampler II probes; one cup of each pair held the sample and the other held the reagent. Stewart (797) modified the Sampler II to hold two 2-ml cups at each position and adjusted the tubing lengths between the sample and reagent probes and the manifold to synchronize the arrival of the two streams at the first point of mixing.

Fasce and Rej (256) reported a different approach. They replaced the Sampler II probe holder with a metal rod 40 cm in length, extending over the sample tray toward the front of the Sampler II. The rod held the sample probe in its usual position; at the end of the rod was mounted a second probe which aspirated reagent from a separate container in synchronization with sample aspiration. A photograph of the rod and probe arrangement appears in their paper (256). By suitably positioning a vessel containing the wash solution for the reagent, the second probe could be made to cycle between the reagent and its wash just as the normal sample probe does. This is one way to economically aspirate a reagent whose wash solution is different from the wash solution required for the samples. Another approach is to install a second wash reservoir next to the standard wash reservoir. Antonis et al. (39) show a photograph of a Sampler II fitted with an extra wash reservoir and an extra probe on the probe arm mechanism. The reagent was placed in a small container (a sample cup) clipped to the front of the extra wash reservoir, and the second probe was mounted so as to dip into the reagent cup as the normal sampling probe dipped into the cup on the sample tray. This system of reagent aspiration was also found to be satisfactory by Zivin and Snarr (940).

Strandjord and Clayson (801) used a Sampler II modified to hold two concentric rings of sample cups. Their paper shows a photograph of the Sampler II, which also featured two probes positioned to aspirate two cups at once on the same axis and a double wash reservoir to supply independent wash solutions to the probes. Several uses for the double ring of cups were described. In one application, the samples were placed in the outer ring, the reagents in the inner ring. The cups in the inner ring alternated between two reagents, and duplicate cups of samples were placed around the outer ring. This arrangement of cups allowed two chemical determinations to be conducted on each sample. In another system, three reagents were used to sequentially test three cups of the same sample. A different application of the double ring of cups permitted determining the blank value of each sample. Every other cup of reagent held a solution from which the key reagent, that is, the reagent producing the sample absorbance, had been omitted. This system of blank determination helped in obtaining more accurate blank values, a feature especially useful when one is confronted with instrument drift. Hopkinson and Lewis (412) published a drawing of a Sampler II modified to aspirate reagent only as samples were aspirated; their modification also involved an additional reservoir for wash solution to service the additional reagent probe.

A further refinement, permitting different sample to wash ratios for sample and reagent probes, was reported by Reed et al. (683). An additional motorized probe drive was installed to aspirate the reagent. The Sampler II was also provided with three cams stacked on one shaft. Additional microswitches were positioned to be tripped by the extra cams. In one procedure, two of the three cams were used to control the aspiration of sample and reagent in proper synchronization. In another procedure, only the third cam was used for normal sampling. The Sampler II alarm switch was rewired to disconnect the two unneeded cam switch assemblies when normal operation was desired.

i. Modified Sample Tray Indexing and Gradients. Several laboratories have modified the sample tray indexing mechanism to permit moving two cups past the probe position in each Sampler II cycle. Tappel and Beck (821) made a special gear for this purpose and also attached a second probe to the Sampler II probe arm. The two probes were positioned so that adjacent cups were aspirated simultaneously. The cups were alternately filled with sample and reagent, and the purpose again was to conserve on the consumption of an expensive reagent. Cacciapuotti (157) also modified the Sampler II to advance the tray two positions at a time and gave a photograph of the special mechanism. The sample tray was modified to hold a second concentric ring of cups. When the standard tray advance mechanism was used (one position indexing per cycle), two probes allowed aspiration of two samples simultaneously to two analytical channels on the manifold. To prepare samples, the special advance mechanism was used to index the sample tray two positions per cycle. Four different reagents were pumped to four probes, two over the inner ring of cups, and two over the outer ring (the cups being prepared in sets of four). One drop of each sample was manually added to each group of four cups and, after sufficient reaction time, the samples were analyzed in pairs using the normal Sampler II advance mechanism, as described above.

Eichler (237) shows photographs of a Sampler II tray mechanism modified to advance the tray two positions per cycle and gives details on how to make the mechanism from standard Sampler II parts. The two-position index feature, plus a total of three probes, made it possible to generate a reagent gradient in the cup adjacent to each sample cup. One of the three probes aspirated the sample cup; the extra two probes dipped into the adjacent "gradient cup." One of these probes delivered reagent from the Pump to the gradient cup, and the second probe removed the gradient liquid from the gradient cup back to the manifold. A mixer stirred the solution in the gradient cup as the reagent was delivered to and from it. A pair of pump tubes, matched within 0.5% in flow rate, was used to move the stream to and from the gradient cup. Later, Fleet et al. (282) reported using Eichler's special tray advance mechanism successfully; they applied it to the aspiration of reagent and sample from adjacent cups through separate probes.

Stansfield and Rossington (786) described another method for producing gradients in the Sampler II cups. Their test required subjecting each sample to a continuously varying concentration of sodium chloride. A double-chamber container was made by putting a tube cap inside of a large sample cup, and three probes were fitted to the Sampler II probe arm. The sample was placed in the tube cap chamber, and the initial concentration of sodium chloride was placed in the outer chamber of the cup. The three probes descended simultaneously. One probe aspirated the sample from its chamber, the second probe aspirated the salt solution from its chamber, and the third probe delivered water, plus air bubbles to help in mixing, into the salt solution chamber, thereby producing the gradient. Stansfield and Rossington (786) give the equations used to calculate the gradient concentration but report that these calculations were inconvenient in processing many samples. They therefore devised a very interesting technique to read out, on the Recorder, a signal which could be calibrated to represent the gradient concentration. A dye, such as phenolphthalein, was added to the solution of sodium chloride initially placed in the gradient chamber. The dye was colorless in the sample stream as it passed through the first Colorimeter, allowing the recording of the sample absorbance only. A second part of the manifold added sodium hydroxide solution to a portion of the stream, developing the color of the dye, and the absorbance of this stream was monitored by a second Colorimeter. Since both the dye and the salt were subjected to the same gradient dilution in the gradient chamber, the absorbance of the dye could be directly calibrated to represent the value of the salt concentration at any given time.

j. Additional Hints on Operation. Multiple probes may be used on the Sampler II to feed streams to multichannel AutoAnalyzer systems. Matusik et al. (582) show a drawing of a Lucite block designed to hold two probes and dip them into the same cup. A metal rod extends from the Lucite block to the Sampler II probe mechanism. Sobocinski and McDevitt (782) give a diagram of a Teflon disk through which nine probes pass; the disk positions the probes to dip into the same cup on a Sampler II.

Many methods of analysis require a knowledge of the sample absorbance with and without addition of a certain reagent to permit subtraction of the sample blank value. Sloman and Panio (778) describe a technique for acquiring these data automatically using the Sampler II. Two probes are arranged to aspirate solutions from two adjacent cups. The cups are filled and placed on the Sampler II tray in the following pattern: a cup of solvent, a cup of sample, a cup of reagent, and a cup of solvent. This pattern of four cups is repeated for each sample or standard around the sample tray. The streams from the two probes are immediately mixed on the manifold, so the sequence of streams entering the manifold is as follows: solvent plus solvent, solvent plus sample, sample plus reagent, reagent plus solvent, repeating with solvent plus solvent, and so forth. With this cup pattern, two peaks per sample (the sample and its blank) are obtained for the calculations. A disadvantage is that the actual sample processing rate is one-fourth of the Sampler II sampling rate.

A technique for rapid sorting of samples showing a positive or negative reaction was described by Hochella and Hill (402, 403). The technique was applied to a rather slow enzymatic reaction, and the reaction was conducted directly in the cups of the Sampler II. The reagent was added manually to each sample in its cup, and the Sampler II was allowed to run through two revolutions of its sample tray, so that each sample was aspirated twice with a time interval between aspirations equal to the time of one tray revolution. The recorded peaks from the first and second run were compared, sample by sample, to reveal suspect specimens.

k. Control of External Devices. The Sampler II has been used to control the operation of various external devices. Dieu (227) analyzed for inorganic sulfates by turbidimetry, as mentioned above in Sec. II, F, 4, and used an air rinsing valve to flush the Colorimeter flowcell between samples. The valve was controlled by the upper portion of a special double cam in the Sampler. The two cams were adjusted to flush the cell with air immediately after each peak was recorded.

Vestergaard and Sayegh (871) described an automatic readout system, using only a Sampler II, a Colorimeter, a Recorder, and a regulated vacuum source. The prepared, reacted samples were placed in the Sampler II. A microswitch controlled by the Sampler II activated a time-delay relay which opened the vacuum line for a timed period, pulling each solution from the

Sampler II into the Colorimeter flowcell. This system was operated at 120 samples per hour.

Manston (568) monitored samples with a Colorimeter and then sent the stream from the flowcell through a second portion of the manifold. After dialysis, the stream was sent to a fraction collector for subsequent radio-isotope counting. A microswitch triggered by a cam mounted above the normal Sampler II cam was used to synchronize the fraction collector to the AutoAnalyzer system.

Berry and Walli (80) built a solenoid-operated automatic microsyringe that was controlled by the Sampler II. The syringe delivered a reagent into the sample cup, starting the reaction just as the probe descended into the sample. The solution was pumped directly from the sample cup through a flowcell, and at a sample rate of 20 per hour the kinetics of each sample reaction could be followed for about 3 min. Instead of a peak height, the slope of the continuous signal, which changed as absorbing material was consumed in the reaction, was used in the calculations. The slope was felt to be more reliable than a single peak height measurement.

Kuzel and Coffey (508, 509) controlled the action of a special cell holder to position the sample flowcell or blank flowcell in the light beam of a Hitachi-Perkin-Elmer Model 139 spectrophotometer by signals from a Sampler II.

Cavatorta et al. (175) used the "final stop" switch on a Sampler II to trip a valve, changing the reagent pulled into a diluent line on the manifold. The Sampler II was operated through two revolutions of its sample tray. The first time around, the samples were aspirated and mixed with a reagent solution, yielding the blank value for each solution. On the second pass, the valve switched a second reagent solution to the manifold, resulting in the color development of each sample.

Neeley et al. (622) used an AutoAnalyzer system with a Coulter Counter. To identify each sample for the Coulter Counter, a valve system was placed in the manifold to momentarily admit a dye solution into the sample stream. The valve was controlled by a microswitch attached to the Sampler II and admitted the dye pulse just as the probe dipped into a cup. A photocell was placed to monitor the stream at the correct point on the manifold and, when the dye pulse reached the photocell, a second two-way valve was triggered to route the stream to the Coulter Counter.

Gray and Owen (345) located the single-pole microswitch inside the Sampler II that stops the probe arm mechanism in the sampling position and replaced it with a two-pole, two-way microswitch. One pole of the new switch acted in place of the original microswitch, and the other pole signalled a computer that the sampling cycle had started or finished. The schematic of this conversion is given in their paper.

We conclude this section by briefly describing the use of the Sampler II as an integral part of four rather complex systems. Tappel and Beck (822) collected the stream exiting from a fluorometer flowcell as 7.7-ml fractions in a Sampler II. This Sampler II then fed the fractions for about 10 min each into an additional AutoAnalyzer system designed to obtain further colorimetric measurements on each collected sample. The Sampler II holding the fractions was controlled by a second (master) Sampler II which contained various reagents and passed them sequentially to the colorimetric manifold. Thus, each sample fraction underwent several tests automatically. The master Sampler II controlled the slave Sampler II by a two-stage cam and microswitch arrangement. Beck and Tappel (68) also described a similar system for multireagent testing. A master Sampler II was fitted with two cams on its drive shaft and was provided with a different cam drive motor which turned the cams at 0.05 rpm. The master Sampler II contained the reagents and, through the action of its extra cam, it controlled a slave Sampler II containing test sample fractions from a chromatographic column. By using the two cams, plus two reset timers, plus various reagents in the cups of the master Sampler II, each even fraction was tested with three reagents, and each odd fraction was tested with three other reagents.

Bradley and Tappel (126) published photographs of sequencing cams which permitted a master Sampler II to control various operations and pictures of a solenoid-operated "uncoupler" that connects or disconnects the probe arm cycling mechanism and the sample tray advancing gear. By using the cams on the master Sampler II (which held samples), the "uncoupler," on-off timers, and a slave Sampler II (which held various reagents), the reactivity profile of each sample could be obtained.

Berry (79) used a Sampler II equipped with a microswitch set to close when the probe arm moved to the sample position. The switch triggered a timer and an Autodilutor device. The Autodilutor pumped the sample from its cup, added reagents, and pumped the prepared sample to a Gilford Model 300-N spectrophotometer. The absorbance was then monitored with time until the timer triggered the Sampler II for the next sample. The Sampler II was fitted with three probes, and the sample reactions were conducted directly in the Sampler II cups by the following scheme: The sample cups are placed in every other position on the Sampler II tray. In the remaining positions, larger cups are placed to act as reaction vessels. The probes descend into the sample cup, and the Autodilutor aspirates the sample through the first probe. The probes return to the wash reservoir and then move to the next (empty) cup. The Autodilutor forces the sample plus a reagent through the same probe into the reaction cup and then pushes air through the second probe to mix the solutions in the cup. After a timed reaction interval, the Autodilutor aspirates the solution through the third probe and passes it to the flowcell for absorbance measurement. Berry gives photographs of the various devices.

3. Sampler IV

Koszyn et al. (496) show pictures of two types of cup for use with the Sampler IV. One type is designed to hold a standard solution, and the other is designed to hold a wash solution. Each type of cup has a trip arm, of different length, which trips the appropriate microswitch on the Sampler IV, notifying a Technicon AutoCal device whenever a standard or wash cup is being aspirated. The AutoCal uses a timer set for the delay between aspiration and the appearance of the peak, and at the proper moment the device automatically resets the blank or standardization circuits on the SMA 12/60 system. A system to warn of excessive drift is included.

4. Sampler 40

Whitehead (905) has published a description, explanation, and pictures of the T-40 Sampler. A punched card sensor system is used to identify each sample; the number punched on the sample card is printed, lined up with the corresponding peak, by the Recorder. The delay between sample aspiration and the appearance of the peak is accounted for by moving the sample probe mechanism so that it samples a cup several positions ahead of the card reader position. Constandse (194, 195) described the IBM card reader system used with the Sampler 40 for specimen identification.

Helmreich (386) discusses a later version of the T-40 Sampler, which optically reads printed labels attached to the specimen cups. The label reader reads the sample identification as the sample is being aspirated rather than several minutes later, as was done with the punched card reader system. The labels may be preprinted or generated by a computer.

Weschler et al. (898) used an IDee Sampler equipped with a special sample probe designed to reduce clogging from blood clots to a minimum. The probe consists of inner and outer concentric tubes. The sample is aspirated through a filter of large surface area, containing many holes of 0.008 in. diameter, and the filtered solution is pumped through the inner tube to the manifold. Between samples, while the probe is in the wash reservoir, a reverse flow of air is forced out through the filter to remove entrapped material. As the probe leaves the wash reservoir, an air bubble serves to partially dry the filter before it enters the next sample. Weschler et al. furnish a drawing of the probe in their paper.

5. Large Sampler

A second probe and sample line have been used to provide a wash from a separate container while the sample probe lifts into the air (585, 787), much the same as described above in Sec. IX,A.

Dewart et al. (226) mention changing the timer on the Large Sampler to bring measurements within ±0.5 sec accuracy, improving the precision of results. Jordan (444) used a Cycl-Flex Timer, made by Bliss-Eagle Signal Company, to obtain the desired precision with the Large Sampler.

Tenny (828) used the Large Sampler in an AutoAnalyzer system for Kjeldahl nitrogen, and adjusted the Sampler to expose the sample line to air for only 2 sec per sample cycle to minimize aspiration of ammonia which was present in the laboratory atmosphere. Tenny also used nitrogen rather than air to mix samples in the Large Sampler module.

Garza and Weissler (310) placed oversized marbles on the sample tubes of a Large Sampler to decrease evaporation of samples. The marbles were automatically removed just prior to sampling.

6. Alternative Sampling Devices

James and Townsend (432) used a sampler made by Hook and Tucker, Ltd., London, with a continuous flow analyzer. The sampler was equipped with two crooks and a constant level reservoir to provide a wash between samples.

Lauwerys et al. (523) used a Carlo Erba continuous flow analyzer. The sampler accepted three concentric rings of sample cups, and the use of these rings to calibrate the analyzer with an internal standard solution is described. First, the samples are placed in the outer ring, and the sampler is operated once around the series of samples. During this run, a key reagent is omitted so as to obtain the sample blank values. Next, the samples are run again, once around, with all reagents supplied, to determine the value of each sample plus its blank. Finally, the internal standard cups, which contain a standard solution of the material being analyzed, are placed in the second ring and, by using two probes, the samples plus internal standards are aspirated and mixed on the manifold. This time around, the value of each sample plus its blank plus its internal standard is recorded. The samples are calculated by the formula

$$C = \frac{R(A_a - A_b)}{A_s - A_a}$$

where C = concentration of sample

A_a = absorbance of sample plus blank

A_b = absorbance of blank

A_s = absorbance of sample plus blank plus internal standard

R = concentration of internal standard

This method of internal standardization was used because differences were noted in the slopes of standard curves obtained by adding known quantities of standards to various samples. By adding standards to each sample, reaction inhibition or accentuation by the sample matrix could be self-compensated.

Henriksen (388) mentioned a specially constructed sampler which provided a wash between samples rather than air. The sampler consisted of an aspirator, turntable, and programmer, electrically interconnected.

XII. SOLIDprep SAMPLER*

A. Descriptions and Notes on Operation

Early prototype versions of the SOLIDprep module are described by Holl and Walton (408) and by Michaels and Sinotte (597).

Beyer (84) and Russo-Alesi (703) corrected irregular sampling by de-bubbling the sample stream from the SOLIDprep and immediately resegmenting the stream with air as it entered the analytical manifold. By doing so, Wrightman et al. (920) obtained more consistent bubble patterns, compared to direct sampling from the SOLIDprep, because the SOLIDprep provided different air to fluid ratios during the sampling and nonsampling portions of its cycle.

Russo-Alesi (703) found that fluid may be forced into the sample line during homogenization. This leakage was prevented by pumping a constant backflow of air until the sampling portion of the cycle started.

One may use a decantation trap fitting in the sampling line from the SOLIDprep to remove most of the solid material before sending the stream to a Continuous Filter module (211). Cullen et al. (207-210) give a drawing of such a decantation trap made from two C1 fittings. Fassari (257) used a series of T fittings in the SOLIDprep outlet sample line, along with a fairly slow pumping rate, to trap insoluble particles in the homogenized samples. The downward arms of the T fittings were capped off. Russo-Alesi (703) used a glass wool plug filter instead of a Continuous Filter module to trap solid material coming from the SOLIDprep.

Ahuja et al. (11, 12) discuss techniques useful in analyzing drug capsules. They recommend loosening the capsules slightly before placing them in the SOLIDprep cups. In one method (11), the SOLIDprep was programmed to begin stirring after all diluent and sample reached the homo-

---

*To the best of my knowledge, all comments in this section are derived from papers reporting work with the original SOLIDprep Sampler, not the recently introduced SOLIDprep Sampler II.

genizer. In a second procedure (12), the best program sequence was found to be (a) a portion of diluent, (b) sample delivery into the homogenizer, (c) the remainder of diluent, and (d) homogenization. Ahuja et al. (12) noted that the drug was not completely dissolved when the SOLIDprep was operated at 20 samples per hour, and used a program rate of 13 samples per hour.

Several useful techniques have appeared concerning the analysis of drug tablets. Bryant et al. (149) presoaked tablets by pumping water at 1.2 ml/min into the SOLIDprep sample cup five positions ahead of the sampling point. Each tablet was soaked in 3.6 ml of water for 12 min prior to sampling, and this softened the hard tablets sufficiently to permit thorough homogenization. Khoury (478) mentions that diatase of malt has been shown to be effective in the desorption of certain steroids from excipients and that this material may be tried in the SOLIDprep when incomplete extraction is suspected. Ahuja et al. (12) used a plastic cover over the SOLIDprep to prevent accidental ejection of tablets during homogenization.

Kuzel (505, 507) used potassium permanganate tablets to evaluate the performance of the SOLIDprep module. The highly colored compound allowed visual tracing of the solution flow. The weight of each tablet was essentially proportional to its potassium permanganate content, and this facilitated calculations during the evaluation.

B. Special Applications

1. Mechanical and Electrical Modifications

Kuzel (505, 507) gives detailed instructions for improving the performance of the SOLIDprep module, including the use of better valves and a sampling device which improves the air bubble pattern and the precision of sample withdrawal. Beyer and Smith (89, 90) confirmed that Kuzel's arrangement of the lines on the SOLIDprep gave better flows and more uniform air segmentation. Stevenson et al. (794) show a Lucite block fitting, designed to be attached to the SOLIDprep, which improved the performance of the module along the lines suggested by Kuzel. They also describe the valve switching sequences during the sampling and nonsampling portions of the SOLIDprep cycle.

Blezard and Fifield (115) gave an alternative approach to improve the performance of the SOLIDprep module. They noted that the SOLIDprep relies on air pressure to prevent the sample from leaking into the analytical manifold during the dilution, homogenization, and rinse portions of the cycle. Air pressure alone was not reliable in preventing such leakage. They devised an external solenoid valve wired to a spare cam switch on the SOLIDprep programmer, which solved the problem. The flow diagrams before and after the conversion are given in their paper.

Mercaldo and Pizzi (592) modified a SOLIDprep module by installing an ultrasonic horn in place of the standard homogenizer. The horn was found to be more efficient than the homogenizer, and it permitted the use of lower volumes of sample diluent. Smith et al. (779, 780) show a picture of a modified blade for the SOLIDprep homogenizer, which improved recovery of soluble potassium from various plant and organic materials.

Hormann et al. (413) changed various components of the SOLIDprep module to achieve efficient extraction of soil samples. Larger cups were provided to hold 40-g samples, and the programmer was modified to yield a 10.5-min cycle. Heating tape was placed around the homogenizer vessel, and the tape was switched on and off at the appropriate points in the cycle to bring the extraction solvent, aqueous acetonitrile, to 50 to 60°C at the end of the blending portion of the cycle. A special solenoid valve was used to add 2.5 to 3 ml of calcium chloride solution to the SOLIDprep homogenizer vessel in order to speed the sedimentation of the soil particles.

Wrightman et al. (920) give pictures and a description of a reservoir which delivers a second diluent into the SOLIDprep homogenizer. The system uses a self-leveling buret and solenoid valve arrangement. A basic, then acidic sequence of solvents was used to dissolve all three ingredients from tablets containing magnesium hydroxide, aluminum hydroxide, and aspirin.

Wachtel and Peterson (878) used 0.1 N hydrochloric acid to dissolve tablets in a SOLIDprep and found that coating the stainless steel homogenizer with Teflon greatly reduced the problem of corrosion.

Siriwardene et al. (765) modified the circuitry of the SOLIDprep module to allow external control of the homogenizer speed. Samples of biological materials were blended before being placed in the SOLIDprep cups, and the homogenizer was operated at low speed to act as a stirrer during the sampling portion of the cycle. The homogenizer was operated at full speed only during the rinsing portion of the cycle.

Beyer and Smith (89, 90) devised a tablet-dispensing apparatus to service the SOLIDprep module. Up to 300 tablets could be loaded in the apparatus, which then dispensed them automatically to the SOLIDprep. A series of pumps was used to prewet the tablets and to deliver standards. A central control unit programmed the operation of these pumps and coordinated them with the SOLIDprep and with the attached tablet-dispensing apparatus.

2. Alternative Solid Sampling Modules

Burns (152) introduced an alternative solid sampler device in which a beaded chain is drawn through a tube containing the sample. The chain pulls the solid sample out of the tube, allowing it to fall into a solvent, and the sample solution thus formed is withdrawn by the AutoAnalyzer Pump. This sampler was used in monitoring the removal of moisture from solid material in a dryer.

Boucher et al. (124) gives pictures and a description of an automatic soil extraction device which prepares the extracts and feeds them to various AutoAnalyzer manifolds.

Chapter 3

THE USE OF LABORATORY INSTRUMENTS
IN CONTINUOUS FLOW ANALYSIS

In addition to the modules supplied by Technicon Corporation for use in AutoAnalyzer systems, a surprising variety of standard or specialized laboratory instrumentation has been used in continuous flow analysis. The purpose of this chapter is to survey the source literature, concentrating on how the various instruments are connected or interfaced to continuous flow analyzers and on how certain problems characteristic of each instrument have been solved.

I. ATOMIC ABSORPTION SPECTROPHOTOMETERS

Atomic absorption spectrophotometers from at least three different companies have been used with AutoAnalyzer systems. Most of the articles listed in Table 1 give explicit details on how the stream from the manifold is sent into the spectrophotometer aspiration-burner assembly. Lacy (512) assembled an atomic absorption spectrophotometer from various manufacturers' components for use with an AutoAnalyzer.

The appreciable vapor pressure of elemental mercury makes possible a gaseous separation of the metal from the flow stream before the mercury vapor is sent to the flowcell in the atomic absorption spectrophotometer. After effecting a chemical reduction of mercury ions to elemental mercury on the manifold, Bailey and Lo (53) added a stream of nitrogen at 120

TABLE 1

Atomic Absorption Spectrophotometers Used with the AutoAnalyzer

Instrument	Reference
Instrumentation Laboratory Model 153	Gochman and Givelber (333)
Perkin-Elmer	
Model 290	Mislan and Elchuk (601, 602)
Model 303	Baker et al. (56)
	Goulden and Afghan (338)
	Lott and Herman (552)
	White et al. (674)
Model 403	Goulden and Afghan (338)
Techtron (Varian)	
Model AA-3	Klein and Kaufman (486)
	Klein et al. (487-490)
Model AA-5	Bailey and Lo (53)

ml/min to the flowing liquid stream. The nitrogen swept the mercury vapor through a gas-liquid separator fitting, which routed the liquid to waste and the vapor to the spectrophotometer flowcell. Goulden and Afghan (338) also reduced mercury ions to metallic mercury on an AutoAnalyzer manifold. The aqueous suspension of mercury was then brought into intimate contact with a large volume of air bubbles by passing the air and the mercury suspension through a mixing coil. A special gas separator fitting then sent the gas phase containing the mercury vapor to the spectrophotometer flowcell. A drawing of the gas separator fitting is given in their paper.

Goulden and Afghan (338) also give details on the construction of a gas flowcell for use with a Perkin-Elmer Model 303 or 403 atomic absorption spectrophotometer. The cell was made from a borosilicate glass tube, 10 mm i.d., and had a 100-mm pathlength. The ends of the tube were ground square, and quartz windows were epoxied on to each end of the tube. The flowcell was mounted on the spectrophotometer burner support, and the spectrophotometer support alignment controls were used to line up the flowcell. A 60-W bulb, shining on the flowcell, maintained it at about $10^\circ C$ above ambient temperature, and prevented moisture condensation in the flowcell.

Bailey and Lo (53) used a Varian Techtron autosampler plus a Proportioning Pump and AutoAnalyzer manifold to prepare samples and send them to a Varian Techtron Model AA5 atomic absorption spectrophotometer. As supplied, the slowest sampling rate of the Varian autosampler module was 25 sec per sample. A rate of 1.1 min per sample was obtained by adding a capacitor to the autosampler timing circuit. A cup of blank solution was placed between each sample cup to permit the Varian "autozero" feature to correct automatically for baseline drift.

Gochman and Givelber (333) determined calcium and magnesium simultaneously using an AutoAnalyzer system and an Instrumentation Laboratory Model 153 atomic absorption spectrophotometer. Magnesium was monitored on the spectrophotometer monochromator channel, and calcium was monitored on the filter channel. The system was operated at 90 samples per hour and required only about 100 μl of sample.

Mislan and Elchuk have described two methods for increasing the sensitivity of AutoAnalyzer/atomic absorption spectrophotometer systems. In the first (602), samples were analyzed for iron, nickel, and chromium in the parts per billion range. A Continuous Digestor was used to concentrate the samples before feeding them to a Perkin-Elmer Model 290 atomic absorption spectrophotometer. Without modifications, the Digestor could be made to yield concentration ratios of the order of 10 to 1. With the described modifications, automatic preconcentrations of about 70 to 1 were achieved. If long periods of time were used (about 3 hr for the system to equilibrate), factors of 250- or 300-fold concentration were feasible. The concentrated sample was collected from the end of the helix into a chamber and, at timed intervals, the concentrate was delivered to the spectrophotometer. In the second method, Mislan and Elchuk (601) used an AutoAnalyzer Proportioning Pump and manifold, along with the same model atomic absorption spectrophotometer. The rest of the apparatus was custom built. An automatic pipet deposited a sample of dissolved salts on a graphite boat. The boat was moved into a purged graphite induction furnace that was heated by copper focusing coils powered at 398 kHz. The salts were converted to atomic vapor and the spectrophotometer viewed the vapor through the axis of the furnace tube. The sample boats were pulled through the furnace-cell area by a graphite transport tape that was powered by a reversible Bodine motor. The sampling system was controlled by a timer, which activated the sample deposit pipet and the tape transport motor. This apparatus permitted analyzing for iron, silver, copper, tin, and aluminum in the submicrogram range.

II. ELECTROCHEMICAL SYSTEMS

A. Amperometry

Bomstein et al. (122) published a report of a continuous flow differential amperometric analyzer. The system used two peristaltic pumps and an associated manifold. No air bubbles were used to segment the streams, but air was used to wash the system between samples. The aspiration tube was placed in each sample manually, and the solution was pumped by the main peristaltic pump through two sample pump lines to the manifold. The two sample streams were passed through separate portions of the manifold, where different reagents were added to each, forming a sample stream and a reference stream. The two streams were then pumped to the differential electrode system, which consisted of two tubular platinum electrodes with associated bridges to a reference saturated calomel electrode (SCE) and exit pathways to waste. A drawing and circuit schematic for this electrode system is given in their paper. The waste liquid was pumped out of the top of twin overflow chambers, open to the atmosphere. The second peristaltic pump controlled the waste liquid outflow rate, which was slightly less than the flow rate into the electrode system. This arrangement pressurized the electrodes slightly and reduced the pulsing effect of the main pump, thus helping to get a smooth recording. The signals from the twin electrodes went through a bridge circuit and then to a recorder. The dual electrode circuit was designed to bridge the blank signal against the sample signal, providing continuous compensation for the value of the blank. The sample rate was about 7 samples per hour. Each cycle consisted of a sample "scrub" of about 2 min, an air wash for 15 sec, aspiration of the sample for 5 min, the taking of the sample reading, an air wash for 15 sec, repeating for the next sample. The precision ran from 0.7 to 2.3%, expressed as a coefficient of variation.

B. Coulometry

Kesler (470) designed a flowcell for use at rates of 2 to 4 ml/min with a Technicon Proportioning Pump. Iodine or bromine solutions of known strength were produced by the accurate measurement of the current flow through the cell.

C. Electrolysis

The design and operation of a continuous flow electrolysis cell was given by Blaedel and Strohl (109). The uses of this cell included the quantitative oxidation of cerium(III) and the separation of mixtures of copper, lead, cadmium, and zinc by electrolysis at various potentials. The cell consisted of a short column packed with crushed graphite, providing an electrode to carry out electrochemistry on the flowing stream. The circuit was completed through a low-resistance solution path. Flow rates up to 6 ml/min were possible using a peristaltic pump. Measurements on the cell effluent were made by absorbance or flowcell polarography, as appropriate.

D. Ion-Selective Electrodes

Cherry (181) gives photographs and descriptions of flowcell electrodes for pH measurements. These assemblies include resistance thermometers to permit temperature compensation. Thompson and Rechnitz (838) show two designs for adapting ion-selective electrodes for measurements in continuous flow streams. These two papers do not concern the AutoAnalyzer, but several of the flowcells seem applicable to continuous flow analysis.

A continuous flow glass electrode used for pH measurement is mentioned but not described by Roodyn (692). Prall (675) gives a drawing of a pH sensor designed for use with the AutoAnalyzer and made from a Beckman combination electrode (Fig. 1). Prall's paper also contains the schematic of a circuit used to connect a standard Technicon Recorder to a Beckman Zeromatic pH meter. Eichler (237) describes how to adapt a standard silver/silver chloride reference electrode and a Beckman Model 39045 glass electrode for pH measurement in an AutoAnalyzer manifold flow stream.

Oliver et al. (635) used an ion-selective electrode in an AutoAnalyzer method for fluoride. After sulfuric acid was added to the sample, hydrofluoric acid was distilled on the manifold. The distillate was mixed with a solution which buffered both pH and ionic strength, and the mixture then was led through a flowcell equipped with an Orion Model 96-09 fluoride electrode. A Thomas Model 4858-L60 pH electrode in a separate flowcell was placed downstream of the fluoride electrode. The pH was monitored to detect variations in the distillation temperature, which affected the amount of sulfuric acid coming over in the distillate. The fluoride electrode response was not linear with the concentration of fluoride ion unless the pH of the flow stream was well buffered, because the pH also influenced the fluoride electrode. The electrode flowcells were constructed from Teflon rods, 1.5 in. in diameter. A well was drilled in the rod to accept the tip of the fluoride electrode. Inlet and outlet holes were provided, and the volume of the flowcell was minimized to improve its washout characteristics. A Swagelok fitting was used to couple the glass bulb of the pH electrode to its Teflon flowcell. Oliver et al. (635) provide a photograph of the electrode flowcells in their paper.

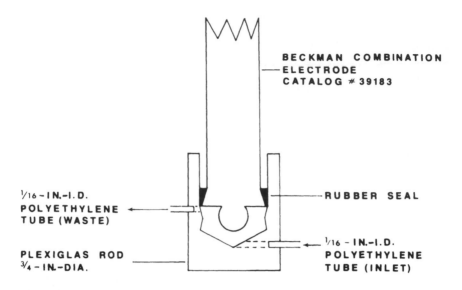

FIG. 1. Prall's adaptation of a Beckman combination pH electrode for use with an AutoAnalyzer. Reproduced from Prall (675) by permission of the author and the Technicon Corporation.

Jacobson (430) determined sodium and potassium simultaneously using Beckman ion-selective electrodes with an AutoAnalyzer. The system noise was lowered by electrically grounding the flow stream. To accomplish this grounding, a platinum wire was inserted just upstream of the electrodes, and it was connected to the "solution ground" terminal on the pH meter. The manifold also used one common reference electrode. In another paper, Jacobson (431) gives examples of manifolds for measuring sodium and potassium using one or two ion-selective electrodes plus a reference electrode. The mathematics used with such electrodes are covered in both papers.

Milham (599) described a flow-through system incorporating an electrode sensitive to nitrate. Although the AutoAnalyzer was not used, the electrode system might be applicable with some modifications to continuous flow analysis. A chamber of about 2 to 3 ml capacity surrounded an Orion Model 92-07 nitrate electrode; a Radiometer Model 601 reference electrode was placed in the exit tube leading the solution out of the measurement chamber to waste. The chamber was fitted with a small magnetic stirring bar. About 30 sec were required for the electrode to reach equilibrium.

Ion-selective electrodes have been used with AutoAnalyzer systems to measure cyanide ion. Fleet and von Storp (283) used an Orion Model 94-06 silver sulfide/silver iodide membrane electrode, and their article shows a drawing of how it was adapted to continuous flow work. A small flowcell with inlet and outlet fittings plus an agar salt bridge to an SCE was attached via an O ring to the tip of the cyanide electrode. The mathematics relating to the mechanism of the electrode response, the influence of pH on its response, and the effect of various interferences are discussed. Variations in electrode output arising from the Proportioning Pump surges were traced to variations in the junction potential of the reference SCE caused by movement of the solution boundary at the porous electrode junction. These variations were suppressed either (a) by using an agar salt bridge along with surge suppressors on the solution line or (b) by pumping the reference electrolyte into the flowing stream just downstream of the cyanide electrode. Cyanide was measured in concentration ranges from 0.01 to 0.00005 $\underline{M}$, at the rate of 40 samples per hour. The limiting factor in sampling rate was the electrode response time, which was about twice as rapid on an upstep change as on a downstep change. The electrode response time was improved by increasing the flow rate through the electrode flowcell. A flow rate of 3.6 ml/min was used in the assays. Llenado and Rechnitz (545) give a figure of a flowcell chamber for an Orion Model 94-06A cyanide electrode used in an AutoAnalyzer system. The sample solution flows into the chamber at a point near the center of the electrode sensing membrane, flows axially outward over the membrane, and then exits near the membrane periphery into a beaker. The reference and solution ground electrodes dip into the solution in this beaker. Typical flow rates through the electrode flowcell were from about 2.5 to 5 ml/min, and the sampling rates were from 20 to 70 samples per hour. An Orion Model 94-53A iodide electrode was used similarly. The reference electrode for either sensing electrode was an Orion Model 90-01.

Zeman et al. (937) described a method for glutamic acid decarboxylase analysis employing a carbon dioxide electrode made by Radiometer Corporation, Copenhagen, Denmark. An AutoAnalyzer system sampled the solutions, mixed reagents into the stream, and sent the debubbled stream to an electrode chamber where pCO_2 was monitored. The electrode chamber was jacketed and was held at $37^{\circ}C$ by fluid from a constant temperature bath. Standard carbon dioxide gas mixtures were fed into the manifold via a T connector to calibrate the system.

Bide and Dorward (92) show a photograph of a Radiometer Model DS66014 flow cuvette which holds a reference electrode plus electrodes for measuring pO_2, pCO_2, and pH. The electrodes worked into a Radiometer Model PHM276M gas monitor and were used with an AutoAnalyzer system for blood analyses.

E. Redox Systems

Sawyer (723) described a redox electrode detection system permitting the analysis of reducing sugars with an AutoAnalyzer. The electrode assembly was made from Technicon fittings, and a figure of the assembly is given in Sawyer's paper. The assembly included a salt bridge to a calomel reference electrode and a platinum wire indicator electrode which projected into the reaction stream. The electrodes were connected to a pH meter monitored by a recorder. The sugars were reduced by ferricyanide reagent, and the resulting ferricyanide-ferrocyanide mixture was monitored by the electrode system. Porter and Sawyer (670) show a figure of a somewhat similar redox electrode flow system, again used with the AutoAnalyzer to analyze for sugars. The flow system fitting includes an integral debubbler. The debubbled solution immediately flows past a platinum indicator electrode and then flows to waste. The fitting also contains a calomel reference electrode (E. I. L. Type RJ 23-1) bridged to the flow stream by a narrow channel of 3 $\underline{M}$ potassium chloride solution. An "earthing electrode" was used to eliminate noise generated by liquid segments moving through the flow lines. A water jacket allowed thermostating of the electrode fitting. The system was operated at 30 samples per hour. Porter and Sawyer (670) also describe how to provide expanded sensitivity and zero adjust for a Pye Dynacap pH meter feeding a Servoscribe RE 511 recorder.

Fleet et al. (281) reported a special porous catalytic silver electrode sensor designed to monitor gaseous oxygen in flowing stream analysis. In one application (281), permanganate or dichromate was analyzed by reaction with excess hydrogen peroxide on an AutoAnalyzer manifold. The liberated oxygen was swept by a flow of nitrogen through two phase separators placed in series. The liquid phase was sent to waste, and the gas phase was pumped through the sensor. Since hydrogen peroxide may be decomposed by excessive amounts of light, black cloth was used to wrap the manifold coils and transmission tubing, as well as the reagent reservoir. In another application Fleet et al. (282) used the silver electrode sensor in an AutoAnalyzer method for chemical oxygen demand. The samples were mixed with reagents and with silver sulfate in 75% sulfuric acid; the mixture was digested as it passed through a Heating Bath set at 145°C. The digested stream was cooled, and the addition of hydrogen peroxide reagent liberated oxygen. The oxygen was swept along by nitrogen, pumped at 6 ml/min, past two gas-liquid separator fittings and then into a U tube, 4 mm i.d., 14 cm in length. Three-quarters of the U tube was filled with soda lime/asbestos (Carbosorb, 12 to 30 mesh with self-indicator) to remove carbon dioxide, and the rest of the U tube was filled with dry calcium chloride to remove water vapor. The gas stream was then passed to the silver electrode sensor device.

F. Polarography

Bertram et al. (81) described a system for "automated polarography" which included a valve, a deaeration chamber, and a dropping mercury electrode (DME). The valve was rotated at 1 rpm, and alternately selected the sample stream and then the supporting electrolyte stream. The streams were sent sequentially to the chamber, where mixing and deaeration were achieved by bubbling nitrogen through the liquid. The prepared solution was then sent to the polarographic cell containing the DME. Solenoid valves were used to drain and refill the polarographic cell for each reading. This cell was not a continuous flowcell, but rather was a stopped flow readout system. Its integration with a continuous flow analyzer seems feasible, however.

Continuous flowcells for polarography have been reported by several laboratories. First, we consider those flowcells not used specifically with the AutoAnalyzer, but which might be applicable. Muller (614) shows a drawing of a flowcell incorporating a platinum wire indicating electrode and a bridge to a reference SCE. The response of the cell was a linear function of concentration and a logarithmic function of the flow rate through the cell. Flow rates of about 0.5 to 5 ml/min were used. Mann (567) used a polarographic flowcell to monitor the eluate from chromatographic columns. The cell was made of borosilicate glass and included a DME and a mercury pool reference electrode. It was operated at about 1 ml/min. Rebertus et al. (682) also used continuous flow polarography to monitor column eluates. They show a deaeration apparatus which removed 95% of the dissolved oxygen from the flow stream. The flow rate was about 5 ml/min, and most of this consisted of the supporting electrolyte which was mixed in just before the stream entered the DME flowcell. Blaedel and Strohl (107) give a figure of a continuous flowcell of about 2 ml capacity featuring a DME. A silver/silver chloride reference electrode was bridged to the flowing solution via a sintered glass frit. Flushing volumes of 2 and 3 ml produced 90 and 99% of steady-state values, respectively. The flowcell was operated with flow rates between 0.5 and 4 ml/min. Later, Blaedel and Strohl (108) reported a different flowcell, made from a Teflon cylinder. A 0.0625-in. hole was drilled axially through the cylinder for the sample stream, which passed by a DME and then by a bridge to a reference electrode. This flowcell could be operated at flows between 1 and 90 ml/min. Scarano et al. (728) give a drawing and description of a continuous polarographic flow-cell designed for flow rates from 1 to 10 ml/min. Their flowcell features a rapid DME mounted horizontally. The mercury stream enters the flow-cell through a tube facing the sample stream entrance tube; the two entrance tubes are separated by about 1 mm in the cell. The mercury drops are thus delivered on the same axis as the sample stream and, since each mercury drop is always surrounded by fresh inflowing sample solution, the response of the cell is very rapid. Low oscillations and very low noise were observed with this cell. An agar bridge to a large reference SCE completed the circuit. Kadish and Hall (450) incorporated a Beckman polarographic oxygen sensor in a continuous flow method for glucose. The addition of glucose oxidase reagent liberated oxygen, and the oxygen content of the flowing stream was monitored by the oxygen sensor and a recorder. Buchanan and Bacon (150) used continuous flow square-wave polarography to monitor the eluate from ion exchange columns. They give a photograph and drawing of the flowcell, which incorporated a DME and reference SCE. At a flow rate of 1 ml/min, a solution of lead in 0.1 $\underline{M}$ hydrochloric acid reached 95% of steady state after 4.8 min and 100% of steady state after 5.8 min. Blaedel and Todd (110) also monitored ion exchange columns with continuous flow polarography. Their flowcell used a DME and featured both a very small holdup volume and very good response times. These authors also described an efficient deaeration apparatus, using nitrogen bubbles, which reduced the oxygen content to less than 0.1% of that of a saturated solution of oxygen in 0.1 $\underline{M}$ potassium chloride.

We next consider polarographic flowcells reported for use with Auto-Analyzer systems. Blaedel and Todd (111) applied their aforementioned flowcell in an AutoAnalyzer method for α-amino acids in column eluates. The acids were passed through a reaction tube packed with 50 to 60 mesh copper phosphate. An equivalent amount of copper(II) was liberated, and the copper(II) was measured in the continuous polarographic flowcell. Lento (529) shows a polarographic flowcell made from a clear Lucite rod, drilled to accept Technicon fittings and a DME (Fig. 2). The flow stream was deaerated by segmenting it with nitrogen bubbles and passing it through a 40-ft delay coil. Lento compares static and flowing polarograms. Fleet et al. (284) give a drawing of a polarographic flowcell used in an Auto-Analyzer method for calcium and magnesium. The cell circuit employs a DME and a normal calomel reference electrode. The flows of various solutions through the cell were controlled by a cam assembly which opened or closed inlet tubes via pushrods. The cam shaft was driven by a gear train connected to the Proportioning Pump motor. Afghan et al. (7) designed a twin cell for a Davis Model A-1660 differential cathode ray polarograph, which allowed the use of this instrument with the AutoAnalyzer. The cells featured the use of an external silver/silver chloride reference electrode that eliminated drift experienced with the conventional mercury pool reference electrode. The cells contained about 5 ml of solution and were made as small as possible without interfering with the drop knock mechanism of the DMEs. A timer was used to control the sequence of events, including the operation of a Technicon Large Sampler module. Fisher electrohosecocks opened and closed the flow of nitrogen from a pressure reduction valve to degas the cells and also controlled the flow of solutions from the AutoAnalyzer Proportioning Pumps. One of these Pumps pumped

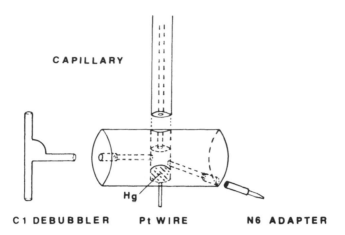

FIG. 2. Lento's polarographic flowcell. The body is a Lucite rod 1.5 in. long, 0.75 in. o.d. The DME capillary is placed vertically through a 0.28125-in. opening. A 0.09375-in.-i.d. hole is drilled on the cylindrical axis and is countersunk to accommodate the C1 debubbler. A 0.09375-in.-i.d. hole is drilled at a 45-deg angle and is countersunk to accept the N6 nipple. A 1-in. length of platinum wire contacts the mercury pool reference electrode and completes the circuit. The solution flows from left to right. Reproduced from Lento (529) with the permission of the author and the Technicon Corporation.

supporting electrolyte to the manifold, and the second Pump pumped samples and standards from the Large Sampler through eight 0.110-in. pump tubes. The sequence proceeds as follows: The sample (or standard) is aspirated at 32 ml/min for 64 sec; then the Sampler probe moves to wash. The valves allow the cells to fill, then drain, then refill. The incoming solution is then valved to waste, and a nitrogen stream degasses the cells. The nitrogen is shut off and the polarogram is recorded. By this time the wash solution is flowing through the manifold, and the cycle repeats, flushing the cells, readying them for the next sample. The system ran at the rate of 15 samples per hour.

III. FLAME PHOTOMETERS

Several laboratories have adapted the Instrumentation Laboratory (IL) Model 143 flame photometer for use with the AutoAnalyzer. Pennacchia et al. (652) analyzed serum samples for sodium and potassium. An Auto-

Analyzer Sampler II was used to feed the samples into an IL Model 144 diluter module attached to an IL Model 143 flame photometer. The Sampler II was operated at a rate of 120 samples per hour with a sample to wash ratio of 1:2. The IL Model 144 diluter was also connected to a supply of lithium sulfate solution, which acted as an internal standard. Amador et al. (23) also analyzed for sodium and potassium in serum samples using an Auto-Analyzer system and an IL Model 143 flame photometer. The flame photometer was equipped with two Duncan 1000 Ω retransmitting potentiometers ganged on the same shaft. One of the retransmitting potentiometers drove the flame photometer readout device, and the other was used to vary a voltage (supplied by a constant voltage source) which drove either a standard laboratory recorder or a Technicon Recorder. Amador et al. (23) furnish schematics for these recorder driving circuits as well as a drawing of an A1 fitting used to debubble the stream and send it to the flame burner. The IL Model 143 aspirator adjustment was used to obtain a flow rate of 1.5 ml/min to the burner. Habig and Williamson (358) used an Auto-Analyzer manifold to add lithium nitrate internal standard solution to samples before feeding them to an IL Model 143 flame photometer. Pulsing from a standard Proportioning Pump I produced fluctuations in the IL Model 143 digital readout device, so they modified the Pump I to run at three times the normal speed. With the modified Pump I, the fluctuations were more rapid, but smaller. The high-speed Pump I also gave a shorter lag time between the introduction of the sample and its readout on the flame photometer, plus a faster rise time to a steady photometer output.

Bold et al. (118) analyzed for sodium, potassium, or calcium in blood samples, employing an AutoAnalyzer system to dilute the samples and feed them to an Eppendorf flame photometer. Their paper shows a figure of a chamber used to debubble the flow stream and deliver a portion of it to the flame photometer burner line. Hurst and Bold (421) modified the Eppendorf flame photometer to allow the simultaneous determination of sodium and potassium with an AutoAnalyzer. Three silicon photocells were mounted behind a 589-nm filter to monitor the flame for sodium. The signal from the silicon cells was sent through a potentiometer to permit adjustment of signal strength and then directly to a 1-mV recorder. The schematic of this circuit is given in their paper. Bold (117) also described the use of these silicon cells in the simultaneous determination of calcium and sodium with the Eppendorf flame photometer and AutoAnalyzer system.

IV. FLUOROMETERS

Vurek (877) describes a circuit designed to stabilize the lamp in an Aminco fluoromicrophotometer. A photodiode is used to monitor the lamp output. A circuit is also given for scale expansion and zero suppression

when the fluorometer is used with a recorder. Vurek's paper may be
of interest when this fluorometer is adapted to continuous flow analysis.
Titus et al. (843) used an Aminco photomultiplier microphotometer connected to a Technicon Recorder via an Aminco adapter (No. 4-8266). Noble
and Campbell (629) describe a modification of an Aminco fluorophotometer
to achieve automatic control over the full-scale sensitivity of the instrument. A microswitch is placed at the 100% end of the Technicon Recorder
scale. When the Recorder pen position exceeds 100%, the circuitry automatically decreases the fluorometer sensitivity, thus keeping peaks on
scale.

Van Dyke and Szustkiewicz (864) describe flowcells which permit the
use of an Aminco-Bowman spectrophotofluorometer (SPF) to monitor an
AutoAnalyzer system.   These authors (865) also give details on how to
drill out the internal adapter for a 1-ml flowcell in the Aminco-Bowman
SPF, so as to pass the maximum amount of light into and out of the flowcell.   Kusner and Herzig (503) connected the Aminco-Bowman SPF to
a Technicon Recorder via an Aminco photomultiplier microphotometer
adapter.

Ott et al. (639) used an Aminco-Keirs spectrophosphorimeter, in the
fluorescence mode, to monitor the stream from an AutoAnalyzer manifold.

An Electronics Instruments Ltd. Model 27A fluorometer was used with
an AutoAnalyzer by Antonis et al. (39).   Their paper gives the circuit
schematic and directions for connecting this fluorometer to a Technicon
Recorder.

The Locarte fluorometer has been used with several AutoAnalyzer
systems.   Flowcells for the fluorometer are described by Merrills (593,
595) and by Martin and Harrison (580).   Merrills (593, 595) connected the
fluorometer to a Technicon Recorder; Martin and Harrison (580) used a
Kent recorder. Tan and Whitehead (820) employed a Locarte Mk 4 fluorometer and described how to wire it to a Technicon Recorder. Muir and Ryan
(611) placed a thallium light source in a Locarte Mk V fluorometer, pre-
sumably to obtain nearly monochromatic excitation energy, and used the
fluorometer to monitor the stream from an AutoAnalyzer.   Rokos et al.
(688) used a Locarte LFM/4 fluorometer with a Technicon Recorder.

A Perkin-Elmer fluorometer was operated as a nephelometer by Savory
et al. (722) in an AutoAnalyzer system. A 360-nm primary filter was used
in the "excitation" train with no secondary filter in the "emission" train.
A square quartz flowcell (Hellma Cells, Inc., No. 176F-QS) was found to
decrease the background signal due to light-scattering, and to provide
optimum optical conditions for nephelometry.   The flowcell wash characteristics required the placement of a wash cup between each sample cup to
eliminate carryover at a Sampler rate of 50 samples per hour, sample to
wash ratio of 1:1, but a timing program of 25 per hour, 1:3, could be used

without the wash cups. Campbell and Gardner (163) published a figure of a debubbler-flowcell assembly used with a Perkin-Elmer Model 203 (or 204) fluorometer at the outlet of an AutoAnalyzer manifold. The assembly was made from a C3 fitting, a short Acidflex tubing sleeve connector, and a short piece of soda glass, 2.8 mm i.d. x 4.9 mm o.d. Blackmore et al. (99) described a simple quartz flowcell for a Perkin-Elmer MPF-2A fluorometer. Their flowcell gave a smoother, less noisy recording than the flowcell supplied with the instrument. A resistor-capacitor network was added at the input of a 10-mV Servoscribe recorder to further suppress noise on the trace.

The Turner Model 111 fluorometer has been used in many continuous flow methods. White et al. (674) describe this fluorometer as used with the AutoAnalyzer. To increase the sensitivity of the Turner Model 111 fluorometer, Fiorica (278) modified a Turner "high-sensitivity unit" to hold a standard Technicon fluorescence flowcell. The modification consisted of a cylindrical aluminum sleeve, 72 mm long, 12 mm o.d., and 8 mm i.d., with four slits, 3 mm x 25 mm, cut at right angles to each other for passage of light into and out of the flowcell. Peuler and Passon (658) used a Turner Model 111 fluorometer equipped with an automated chemistry door (Turner No. 110-690). The debubbler (Turner No. 110-692) inside the door was modified by cutting off the debubbler arm and removing the fritted glass disk from inside the flowcell. To allow the pumping of fairly concentrated solutions of hydrochloric acid through the flowcell, the original non-acid-resistant waste line fitting was replaced with a polypropylene fitting. Beyer (85) employed a Turner Model 111 fluorometer as a continuous flow nephelometer. The fluorometer was equipped with a blue lamp. Wratten 2A-12 filters were placed in both the primary and secondary trains, and the instrument monitored the flowing stream at about 510 nm. A square flowcell and a masked adapter insert were used (Aminco Nos. B16-63019 and A363-62140). Ko and Royer (494, 699) used an Aminco-Bowman cylindrical flowcell of 2 mm i.d. in a Turner Model 111 fluorometer. Mirrors were placed to reflect the incident light passed through the solution back into the solution, and to reflect back toward the detector the fluorescent radiation originally emitted away from it. The use of these mirrors gave more than a twofold increase in sensitivity.

V. GAS CHROMATOGRAPHS

The AutoAnalyzer has been used to prepare samples for subsequent analysis by gas chromatography. Ek et al. (239) described a manifold which extracted methapyrilene and pyrilamine into chloroform and passed a portion of the chloroform extract to an automatic fraction collector. The samples and standards contained an internal standard which passed through the same extraction and collection process. Hormann et al. (413) used a SOLIDprep Sampler, operating at 10.5 min per sample, to extract herbicides from soil samples into a mixture of acetonitrile and water. The extracts were then passed through the helix of a Continuous Digestor module to remove the acetonitrile. The remaining aqueous phase was collected and automatically extracted with a mixture of hexane and ether. The organic extracts were separated and delivered to a Hosli Model V 40 S fraction collector, which was synchronized with the AutoAnalyzer system. The tubes in the fraction collector were held at 80°C to achieve simultaneous collection and concentration of the samples by partial evaporation of the organic solvent. The volume of each concentrate was manually brought to 2 ml, and each extract was manually placed in an automatic gas chromatograph injection system (Hewlett-Packard Model 7570A) for determination of the herbicides. A program using a Wang 700 calculator to process the results is also described.

Stockwell and Sawyer (799) reported a totally automated gas chromatographic determination of alcohol in tinctures and essences. The samples and standards were placed in 15-ml test tubes and were sealed with Parafilm. The test tubes were loaded into an "autosampler" (marketed by Evans Electroselenium Ltd., Halstead, Essex, England). The autosampler probe pierced the Parafilm seal, and the sample was pumped by a Technicon Proportioning Pump to the manifold, where the sample was mixed with an internal standard solution. The stream was then led to a special injection vessel which comprised the actual interface to the gas chromatograph. This vessel collected the prepared solution, and it was automatically drained several times to rinse out the preceding sample. The injection vessel was fitted with a capillary tube connected to the injection port of the gas chromatograph. At the appropriate point in the cycle, a program timer opened a gas valve for about 2 sec, pressurizing the injection vessel and forcing a small portion of the accumulated sample solution through the capillary tube into the injection port of the gas chromatograph. When replicate standards were analyzed, the precision of the ratio of alcohol to internal standard, expressed as a relative standard deviation, was well below 1%.

VI. RADIOACTIVITY

A. Gamma Ray Spectrometry

Briscoe et al. (134) describe a soda glass flowcell coil which was used in a single-channel gamma ray spectrometer (9000 Series, Research Electronics, Ltd., England). The spectrometer and flowcell monitored the stream from an AutoAnalyzer manifold, permitting the analysis of traces of mercury by isotope dilution.

B. Geiger Counter

Blaedel and Olsen (105) gave details on the construction of an all-Teflon flowcell for use with an end-window Geiger counter. A spiral groove was cut in a Teflon block, and the block was provided with entrance and exit fittings. A thin sheet of Teflon was sealed over the spiral to serve as a window. The internal volume of the flowcell was 0.4 ml and it was used with flow rates up to 10 ml/min.

C. Scintillation Counters

Several laboratories have reported flowcells for scintillation counting which, while not used with AutoAnalyzer systems, seem applicable for such use. Tkachuk (844) designed a scintillation detector for monitoring column effluents or other aqueous solutions. Three holes were drilled in a Lucite block and were interconnected so that the solution flowed sequentially through them. The holes were packed with 100 to 200 mesh plastic phosphors, and the assembled flowcell was mounted on an EMI multiplier tube of 2 in. diameter. The flowcell and multiplier tube were placed inside a lead castle, 2 in. thick, and black tape was used to keep stray light from striking the detector. The multiplier tube was water cooled to achieve constant temperature. The flowcell volume was 1.02 ml and it was operated at about 0.5 ml/min. McClintock et al. (583) show a drawing of a modified Packard flowcell used in a Packard Model 3314 liquid scintillation counter. The flowcell was packed with anthracene crystals; its internal liquid capacity was 4 ml when empty and 1.2 ml when packed with the crystals. Schutte (745) gives descriptions and drawings of two types of flowcell for scintillation counters used to monitor the effluent from columns. The first type, heterogeneous, was a U-shaped coil packed with cerium-activated lithium glass beads or other materials. The second type, homogeneous, was a helical coil of 2 mm i.d. and 1.4 ml volume. The flowcells were operated at 2.3 ml/min and were designed for use in a Tracerlab Coruflow Model SCE-542 scintillation counter.

We now turn to the use of scintillation counters with AutoAnalyzer systems. Pollard et al. (668) added $[^{131}I]$triiodothyronine to serum samples. The unbound reagent was separated by dialysis and was then measured by passing the stream through a Teflon coil mounted in the well of an EKCO Model N664A scintillation counter. A description of the flow coil is given in their paper. This Teflon coil and scintillation counter arrangement was also used by Garnett et al. (307). In another paper, Pollard et al. (667) mention that the counter was coupled through a rate meter to a modified Technicon Recorder. Webber et al. (893) also reported using a flow stream coil mounted in the well of a scintillation counter; the counter was connected through a rate meter to a Technicon Recorder modified to respond as a 100-mV potentiometric recorder. Stegink (790) used an Auto-Analyzer manifold to monitor column effluents. A portion of the stream was sent through a flowcell of 1 ml capacity packed with anthracene (Nuclear Chicago 6900, 6787), and the flowcell was placed in a Nuclear Chicago Unilux II scintillation counter.

Ruzicka and Williams (709) applied the AutoAnalyzer to continuous substoichiometric analysis. In this technique, a solution of a radioactive isotope of the element to be analyzed is pumped into the manifold at a constant rate. It is added to the sample stream, and the mixture is sent through a separation process, for example, an extraction. After the separation step, the stream is passed through a radioactivity counter. Thus, the "baseline" is a strong, constant response from the counter. The separation step is purposefully adjusted to be substoichiometric; that is, less than 100% of the element is recovered. As the nonradioactive element from a sample enters the stream, a portion of it passes through the separation step. The amount of the radioactive isotope recovered is thereby further reduced, and a negative peak is recorded. The technique is rather sensitive, in the range of milligrams to nanograms per milliliter.

Ruzicka and Lamm (708) described the analysis of small amounts of mercury by substoichiometric separation and radiochemical analysis. The mercury from the sample and radioactive ^{203}Hg from the reagent were combined and extracted into a solution of dithizone in carbon tetrachloride on the AutoAnalyzer manifold. The stream was then monitored by a scintillation counter. The flowcell device in the counter was made from 3.5-mm-o.d. glass tubing and it included an integral debubbler-segmenter fitting. The mixture of air, aqueous, and organic phases entered the device, which separated the organic layer and discarded the air and aqueous phases. The organic layer was then resegmented with a nonradioactive aqueous solution, 0.1 $\underline{M}$ in hydrochloric acid and 0.5 $\underline{M}$ in acetic acid. The segmented organic stream then passed through the flowcell itself. The segmenting solution helped to clean the flowcell as well as to improve its wash characteristics. A Frieseke and Hoepfner scintillation counter equipped with a well-type NaI (Tl) scintillation detector was used. The counter was coupled through a rate meter (Frieseke and Hoepfner Model FH57B) to a recorder (Elliot Dynamaster Model 570), and the authors state that the circuit diagram for this arrangement may be obtained from them on request.

In another paper on automated trace analysis by radioactive isotopic dilution, Ruzicka and Lamm (707) describe a flowcell made from A1 and B1 fittings plus a glass spiral, of 0.9 ml volume, made of 3.5-mm-o.d. soda glass tubing and used in the well of a Frieseke and Hoepfner scintillation counter.

Dufresne and Gitelman (233) reported an interesting and different use of a scintillation counter in an AutoAnalyzer method for adenosine triphosphate. The method employs luciferin-luciferase reagent, which emits light in the reaction. The light was monitored by passing the glowing stream through 3-mm Pyrex glass tubing formed into a coil of 0.5 in. diameter and 1.5 in. length. This flowcell was placed in a Packard Model 317 flow detector connected to a Packard Tri-Carb Model 314 F liquid scintillation spectrometer.

D. Special Systems

Goldstein et al. (334) used an AutoAnalyzer manifold to perform preparative chemistry on fractions from column eluates sampled by a Sampler II. After passing through the manifold, the samples were automatically deposited by a specially built probe mechanism onto paper disks held in a custom-built wheel. The sample deposition system was synchronized to the AutoAnalyzer by the signal from a Colorimeter. The Colorimeter was arranged to monitor the stream for a dye solution continuously pumped to the Sampler II wash box. A rise in transmittance indicated that the wash solution was leaving the Colorimeter flowcell and that a sample was appearing in the stream. This rise in transmittance triggered the timed deposition of sample from the flow stream onto the paper disk in the output sample collection device. The disks were subsequently placed in a scintillation counter, and the readings allowed the plotting of chromatograms of the radioactive materials.

Bagshawe et al. (52) performed radioimmunoassays of human chorionic gonadotropin and lutinizing hormone with components of an Auto-Analyzer system, plus a specially built "continuous filter" apparatus, using glass fiber paper and supporting filtration paper to collect the radioactive species. From the supply spools, the paper passed under a "presoak" delivery head, received the sample, passed under two wash streams, moved past a hot air blower, and then through two paired scintillation counters. Pollard and Waldron (666) used an AutoAnalyzer system in a radioimmunoassay for insulin. The radioactive constituent was collected by a custom-built "continuous filter" module. Whatman glass fiber paper was first strengthened by applying a stream of rubber solution at each edge just before it passed over the filtration (suction) block. After receiving the sample, the paper was dried and then overlaid with adhesive cellulose tape for further strength. It then passed through a continuous radioactivity scanner.

VII. REFRACTIVE INDEX

Marsh and Fingerhut (572) used a Waters Associates differential refractometer with an AutoAnalyzer manifold to determine serum protein. The refractometer response was linear with concentration, and it was capable of reading 0.00006 refractive index units full scale. The line pressure of the sample flow stream and the temperature of the refractometer sample and reference cells were critical factors in obtaining correct response. The sample line pressure was controlled by a vent to the atmosphere just ahead of the refractometer sample cell. The vent also debubbled the sample line. The sample flow stream was preheated to about $50^{\circ}C$, the temperature of the refractometer, before being sent into the flowcell.

VIII. SPECTROPHOTOMETERS

A. Infrared

Ellis (243) shows a photograph of a flowcell designed for a Perkin-Elmer Model 21 infrared (ir) spectrophotometer. The flowcell was made from a standard demountable 0.2-mm cell by attaching tubing to its upper and lower ports. The flowcell was used manually, samples being introduced with a syringe. Two milliliters of solution was sufficient to flush the flowcell.

Robbins (686) used a Perkin-Elmer Model 237 ir spectrophotometer with an external recorder attachment and described how to use demountable cells, of 3 mm pathlength with Irtran-2 windows, as flowcells with an Auto-Analyzer manifold. Ryan et al. (710) constructed a "minimum volume" flowcell of 3 mm pathlength for use in a Perkin-Elmer Model 521 ir spectrophotometer set at 1728 cm^{-1}. The flowcell was made from a standard sodium chloride demountable cell. A lead spacer was cast around two LuerLok syringe needles, to provide inlet and outlet fittings. The spacer contained a 2.5-mm x 9-mm aperture for the light beam. The completed flowcell was sealed by amalgamation. Its internal volume was about 0.7 ml. A photograph of the flowcell appears in their paper. Ryan et al. (710) also gave a useful technique which reduced etching of the flowcell windows. The AutoAnalyzer manifold extracted the material to be analyzed from an aqueous phase into a chloroform stream. The "wet" chloroform stream was then diluted with dry chloroform before being sent to the flowcell. Urbanyi and Lin (850, 851) used a barium fluoride 1-mm flowcell in a

Perkin-Elmer Model 621 ir spectrophotometer set at 1730 cm^{-1} and at 10x scale expansion. The AutoAnalyzer manifold extracted the analyte into chloroform and then pulled the chloroform stream through the flowcell. A cell filled with chloroform was placed in the reference beam.

B. Ultraviolet-Visible

Before we discuss specific instrumentation, several general areas deserve attention. Roudebush (698) reported that, when Acidflex tubing is used to pump chloroform, materials may be extracted into the solvent which interfere with measurements in the uv portion of the spectrum. Small vapor bubbles may also form as the chloroform leaves the pump tube. Roudebush used displacement pumping to get around these problems. Kuzel and Coffey (508, 509) found that uv interferences were picked up from Tygon tubing, especially when used with warm alkali. They recommend the use of Teflon, polyethylene, or glass tubing wherever possible on the manifold. Jamieson and MacKinnon (434) mention that uv interferences may be extracted into isooctane from Solvaflex pump tubes.

Anderson et al. (31, 32) reduced noise on the recorded trace resulting from schlieren in the flowcell by taping a single thickness of glassine weighing paper with Scotch Magic Transparent Tape over the flowcell window on the side nearest the phototube. Beyer (87) also used this idea successfully. Other preventative measures to reduce schlieren are discussed in Chapter 2, Sec. II, B, 4, c.

Kuzel and Roudebush (510) used potassium chromate solution to evaluate various flowcells at rates of 30 and 60 samples per hour with a sample to wash ratio of 2:1. They give many figures of the recorded peak shapes obtained when using Beckman and Hitachi spectrophotometers with Auto-Analyzer manifolds.

It is often desired to extract the analyte from an aqueous solution into an organic phase on the AutoAnalyzer manifold and then to monitor the organic phase in a spectrophotometer. In such cases, traces of water in the flowcell may lead to noisy recordings. Robertson et al. (687) eliminated water droplets from a flowcell by removing the inlet line between the debubbler and the flowcell, pinching off the line so that all fluids went to waste through the debubbler. Two to three milliliters of ethanol were injected into the flowcell with a syringe, and the inlet line was then reconnected.

Several laboratories have published continuous flow methods in which the manifold includes two distinct channels, one feeding a sample flowcell and the other feeding a blank or reference flowcell. There are two reasons for using this approach: automatic compensation for the blank of each sample (50, 216, 366, 789) or the aging of a reagent during the day's run (559, 742). Ayers (50) steam-distilled nicotine alkaloids on an AutoAnalyzer manifold and then split the distillate into two streams. Hydrochloric acid, which intensified the alkaloid uv absorbance, was added to one of the streams, and this stream was sent to the sample flowcell. An equal amount of water was added to the other stream which was then sent to the reference flowcell. Hall and Meyer (366) extracted barbiturates into chloroform, then into a pH 10 aqueous borate buffer, on an AutoAnalyzer manifold. The borate buffer solution containing the barbiturate was then split; one portion was diluted with the pH 10 buffer, the other with aqueous hydrochloric acid. The alkaline stream was passed through the sample flowcell and its corresponding acidic stream was passed through the reference flowcell. The barbiturate uv absorbance was much lower in acid than in base, and the blank of each sample was thus automatically "referenced out" by the spectrophotometer. Steele and Mansdorfer (789) fed the two streams from an AutoAnalyzer manifold to two separate spectrophotometers set at the same wavelength. They noted that transmission tubing should be kept away from heat sources to avoid differently shaped absorbance peaks from the two channels.

We now review techniques and applications associated with specific instruments. Photometers are included in this section because of their operational similarity to spectrophotometers.

An Aminco microphotometer was used by Van Dyke et al. (862, 863) in an AutoAnalyzer method for adenosine triphosphate determination. The photometer measured the light released in the oxidation of luciferin (firefly extract) by molecular oxygen in the presence of adenosine triphosphate and magnesium ions. The reaction stream was passed through a small mixing coil placed in front of a photomultiplier tube, and a mirror was placed to the rear of the coil to reflect as much light as possible into the photomultiplier tube. Figures of the apparatus are given in their paper. This detection system had a very wide dynamic range (0.1 mg/ml to 1 ng/ml), which was controlled by the gain setting of the photometer circuit.

Shepp et al. (761) and Bomstein et al. (121) show a flowcell which adapts a Bausch & Lomb Spectronic 20 colorimeter for use with an Auto-Analyzer. Bomstein et al. (120) have also described how to adapt and use a Technicon 10-mm flowcell in this instrument.

Several other papers may be of interest when considering the use of the Bausch & Lomb Spectronic 20 in continuous flow analysis. Creamer (201) increased the stability of the instrument by replacing certain resistors with others having higher heat dissipation ratings. The replacement resistors fit onto the original circuit boards, and they reduced the instrument warmup time from about 90 min to about 30 min. Pardue and Deming (646) substituted an operational amplifier circuit for the Spectronic 20 electronics. The source lamp stability was improved by using optical feedback. With these modifications, the instrument was usable in the range of 0.005 absorbance full scale. Bidlingmeyer and Deming (93) also described an operational amplifier circuit which improved the stability of the Spectronic 20 and allowed range expansion of signals from the instrument. These authors used a flowcell in the instrument to monitor chromatographic column eluates. Roubal (696) gives drawings and directions for making flowcells for the Spectronic 20. The flowcell design emphasizes good wash and prevention of bubble holdup. Tubing of 2 mm i.d. is used, and the pathlength is chosen as desired. Epps and Austin (244) give figures of a flowcell for a Spectronic 20 and of a shutter mechanism for balancing the instrument. The flowcell was made from 4- and 10-mm lengths of Pyrex tubing and it was filled manually.

Schwartz et al. (751) used a 1-cm-pathlength flow cuvette in a Technicon Colorimeter at 680 nm and noted that the Colorimeter gave absorbances about 75% of those found when the same cuvette was used in a Beckman DU which had been modified to hold it. They ascribe this difference to the use of filters in the Colorimeter versus monochromatic light in the DU. Zak et al. (932, 933, 935) adapted a Beckman DU for use with an AutoAnalyzer manifold. A logarithmic converter (Ledland Instrumental Engineering Co., Detroit, Michigan) permitted the direct recording of absorbance. A Technicon flowcell was used (932, 935). The flowcell was taped in the cell compartment so that its optical flat area covered the monochromator exit slit (933).

Anderson (30) describes quartz flowcells of 0.2 and 1 cm pathlengths for use in a Beckman DU. Figures are given for the flowcell holders and for a light-tight lid closure containing inlet and outlet flow lines for use in the DU cell compartment. Niebergall and Goyan (626) made a flowcell for the DU from 5-mm-i.d. silica glass bent into a U-shaped tube. Two holes were drilled in the DU cell compartment cover; the tube was fitted into these holes and its protruding ends were painted black. A Beckman Model 5800 energy recording adapter was used with a standard laboratory recorder. Marr and Marcus (569) give a circuit which yields a voltage linear with absorbance from a Beckman DU. The circuit is designed to drive a 50-mV recorder. Rasmussen and Nielsen (681) reported a circuit using operational amplifiers to improve the stability of a Beckman DU and to provide a recorder output.

Beckman DB and DB-G spectrophotometers have been used by many laboratories in continuous flow analysis systems (216, 366, 559, 742, 789, 911). Roudebush (698) evaluated the performance of two flowcells in the DB as used with AutoAnalyzer manifolds. Ayers (50) shows a figure of a 1-cm flowcell for a Beckman DB, usable in the ultraviolet and visible ranges. The flowcell body was made of glass, with silica windows cemented in place. Jamieson and MacKinnon (434) used a Technicon quartz tubular flowcell in a DB set at 275 nm and found that a deuterium lamp gave sufficient energy to operate the system satisfactorily but that a hydrogen lamp did not. Ahuja et al. (12) reduced instrument noise by placing an aperture of 2 mm diameter before the spacer block on the reference side of a DB. Khoury and Cali (479) used a Technicon control timer and a DB-G equipped with a wavelength programmer accessory to obtain automatic repetitive spectral scanning. The timing of the system was adjusted to allow each sample to come to steady state, and the DB-G then recorded the spectrum of each sample over a preselected wavelength range.

Roubal (695) gives pictures, drawings, and instructions on the construction of a quartz window flowcell, of 0.3 ml volume and about 0.08 in. pathlength, for use in a Beckman DB or similar instrument. The flowcell is designed to give good wash, to reduce "channeling," and to avoid entrapment of air bubbles. The flowcell could be used at rather low flow rates (4 ml/hr). Hunt (419) modified a commercially available silica flowcell (Scientific Cell Company, No. 482) of 2 mm pathlength for use in a DB-G to monitor the uv transmittance of column eluates. Hunt also describes how to change the spectrophotometer output voltage from 100 mV to 1 V at 100% T and outlines the use of operational amplifier circuits to get linear absorbance ranges of 0 to 0.25, 0 to 0.5, and 0 to 1 absorbance.

Alber et al. (14) modified a Beckman DK2-A spectrophotometer to permit continuously monitoring the stream from an AutoAnalyzer manifold at constant wavelength. They also describe the interfacing of the DK2-A to a digital computer.

Shapira and Wilson (757) published directions for fabricating micro-flowcells of 10 mm pathlength and about $72 \mu l$ internal volume operable at flow rates of about 2 ml/min. The flowcells were made from drilled and reamed black Plexiglas blocks and were fitted with quartz windows. Wilson (914) used such a flowcell in a Cary Model 14 spectrophotometer.

Robertson et al. (687) used an AutoAnalyzer manifold to extract drugs, as their ion pairs with bromthymol blue or bromcresol purple, from an aqueous phase into a chloroform phase. The absorbance of the chloroform phase was then recorded. Various flowcells were compared as to their susceptibility to interferences from traces of the aqueous phase in the flow stream. The Technicon 15-mm tubular flowcell, operated in a Technicon Colorimeter, was satisfactory if kept completely free of traces of water. However, flowcells with parallel faces, either of a cell width greater than the light beam width or operated with an aperture, were much less troublesome. Three such flowcells were evaluated in a Coleman Model 111 spectrophotometer at a flow rate of 1.6 ml/min. A flowcell of 0.25 ml capacity (Arthur H. Thomas, No. 9120-N05) was more susceptible to traces of water

than was one of 0.4 ml capacity (Fisher, No. 14-385-926B). A flowcell of
0.6 ml capacity (Fisher, No. 14-385-926A) gave incomplete wash between
samples.

In 1961, Wood and Gilford (918) described a device which was designed
as an attachment to a Beckman DU monochromator; it provided an electrical
signal linear with absorbance that was suitable for driving standard laboratory recorders. The schematic of the circuit is given in their paper. This
device increased the long-term stability of the readings from the DU. The
"Gilford attachment" for the DU is commercially available, and it represents an attractive option for a laboratory desiring to convert a Beckman
DU for use in continuous flow analysis. For example, Gawthorne et al.
(313) used a Beckman DU-Gilford system as a "colorimeter" with an Auto-
Analyzer manifold.

Gilford Instrument Laboratories has introduced several spectrophotometer systems, some of which have been used with AutoAnalyzer systems.
Siggaard-Andersen and Oliver (762) give figures and a description of an
adapter which allows use of a Technicon 15-mm flowcell (from a Colorimeter I) in a Gilford Model 300 spectrophotometer. The adapter is made of
aluminum and has entrance and exit ports through which water may be
circulated to hold the flowcell at a constant temperature. White and Gauger
(902) give a schematic of a circuit used to interface the Gilford Model 300
spectrophotometer to a Technicon Recorder. Neeley et al. (621, 623) used
a Gilford Model 300N spectrophotometer driving a Heath IR-18M recorder
to monitor the stream from an AutoAnalyzer manifold and gave the schematic of a low-pass filter which damped the signal to the recorder. A
potentiometer, set at 8000 Ω, was placed in series with the positive lead
to the recorder, and it was followed by a 100-μF, 6-V capacitor placed
across the recorder inputs. A similar circuit was reported by Amador
and Urban (24). It was designed to damp the signal to a Gilford Model 242
recorder, and a 4000 Ω resistor was used instead of the 8000 Ω potentiometer. Yee and Zin (924) compare curves from an AutoAnalyzer manifold
measured at 30 samples per hour with a Technicon Colorimeter and at
50 samples per hour with a Gilford 300N spectrophotometer. They found
that the Gilford flowcell possessed better washout characteristics than the
Technicon flowcell. Hellerstein et al. (385) operated an AutoAnalyzer
manifold with a Gilford Model 300N spectrophotometer and a Gilford Model
4006 Data Lister. The Data Lister was provided with a switch manually
depressed to activate the data printer. To automate the data acquisition
step, a synchronous timer was arranged to trip the switch at regular
intervals. The timer was set to print a reading every 30 sec on the
baseline and was then set to print a reading every 2 sec when the peaks
began coming out. The printer tape was scanned visually to locate the
maximum absorbances representing the value of each peak. Berry (79) used
a Gilford Model 300N spectrophotometer, a Heath recorder, an "Autodilutor"

device, and AutoAnalyzer components in a special system to perform kinetic assays of serum enzymes. This system is described in Chapter 2, Sec. XI,D,2,k. Huemer and Lee (418) employed a Gilford Model 2000 recording spectrophotometer set at 750 nm to monitor the stream from an AutoAnalyzer manifold. They show a photograph of a flowcell made to their specifications by Precision Cells, Inc., New York, of 10 mm pathlength and 0.2 ml volume. The flowcell was designed to minimize turbulent flow and to eliminate air trapping. Huemer and Lee also found that a standard Gilford flowcell worked satisfactorily. The spectrophotometer could be operated at 0 to 0.2 absorbance full scale. Strandjord and Clayson (801) reported an AutoAnalyzer system incorporating a Beckman DU set at 340 nm, a Beckman flowcell, and a Gilford Model 2000 multiple-sample absorbance recorder. Pitot and Pries (660) used a system comprised of a Gilford Model 2000 multiple-sample system, a Beckman DU monochromator, a Technicon Large Sampler, and a Technicon Proportioning Pump. The system sampled and read up to four solutions on a timed cycle by means of a multiple flowcell and indexer mechanism. The Proportioning Pump was switched off during the absorbance measurements.

A Hitachi Model 101-UV-VIS spectrophotometer was used by Looye (547) with an AutoAnalyzer manifold. Looye adapted the spectrophotometer for use with the Technicon Recorder by providing a reference voltage signal to simulate that normally supplied by the Technicon Colorimeter. The reference voltage was obtained from a 1.5-V battery, and it was adjustable from 0 to 12 mV. Beyer (87) and Anderson et al. (31, 32) used a Hitachi-Perkin-Elmer Model 139 spectrophotometer equipped with an Arthur H. Thomas flowcell to monitor the stream from an AutoAnalyzer manifold. Kuzel and Coffey (508, 509) reported an interesting adaptation of the Model 139 spectrophotometer which permitted automatic recording of sample and blank absorbances. The sample stream was fed through two appropriate manifolds, providing sample and blank flow streams. Two flowcells were held in a specially built cell positioner mechanism in the Model 139, and each stream was passed through its respective flowcell. The flowcell positioner mechanism was controlled by a modified Sampler II module, and the sample and reference flowcells were alternately positioned in the spectrophotometer light beam. After proper synchronization of the flow streams with the action of the flowcell positioner, each blank value was recorded sequentially with each sample value on the same recorder.

Conway and Lethco (196) described a flowcell holder for a Photovolt Model 525 densitometer. The holder was used with standard rectangular 1-cm flowcells. A circuit is given to attenuate the output of the Photovolt densitometer for use with standard laboratory recorders. Oleniacz et al. (634) built an AutoAnalyzer manifold to detect microorganisms by chemiluminescence using Luminol reagent. They described a specially designed flowcell mounted in front of a photomultiplier tube connected to a Photovolt photometer. The flowcell and photomultiplier arrangement measured the light emitted in the flow stream.

Pictures and a description of how to modify a Technicon flowcell holder to fit a Turner Model 330 spectrophotometer are given by Krieg and Hutchinson (500). These authors also give a schematic of a constant voltage source which adapts the Model 330 output to drive a Technicon Recorder.

A Unicam Model SP800 recording spectrophotometer provided complete uv spectra of drugs in urine samples in conjunction with an AutoAnalyzer system reported by Blackmore et al. (99). Their manifold was operated at 10 samples per hour, and the spectrophotometer scans were synchronized with the action of the Sampler II. Vandermeers et al. (860) presented a general method for kinetic measurement of enzymatic activities which used an AutoAnalyzer system feeding solutions to a Unicam Model SP800 spectrophotometer equipped with an LKB 4712 A-4 flowcell. The samples were aspirated by a Large Sampler, mixed with reagents, and incubated in jacketed mixing coils on the manifold. They were then read in the spectrophotometer. The overall system was controlled by a timer.

Crestfield (204) describes a microquartz flowcell for use in a Zeiss PMQ II spectrophotometer. The flowcell was used to monitor eluates from columns and was operated with flow rates of about 0.5 ml/min. Crestfield gives a drawing of the flowcell and mentions that a similar flowcell is available from Pyrocell Manufacturing Company, New York.

C. Special Photometers

Blaedel and Hicks (103) describe in detail a filter photometer used in a continuous flow analysis of glucose. After addition of the reagents, the sample stream flowed through two flowcells in the photometer. A delay coil placed between the flowcells permitted kinetic measurements to be made on the reaction stream. Blaedel and Hicks (102) also reported the use of this filter photometer in a continuous flow analysis for lactic dehydrogenase in blood serum. After the reagents had been added, the sample stream passed through a 10-ft length of 0.05-in. plastic tubing in a $37^{o}C$ bath, through the first flowcell, through 5 ft of additional tubing in the bath, through the second flowcell, and then to waste. The delay between the flowcells was about 25 sec, and the difference in the recorded absorbances was a measure of the lactic dehydrogenase present in the serum samples.

Pardue and Rodriguez (647) designed a highly stable photometer used in the range of 0.01 to 0.34 absorbance for kinetic studies. The photometer made use of optical feedback by monitoring its source lamp output and correcting for changes in the lamp's brightness by controlling the lamp's power supply. The schematic for the operational amplifier circuitry is given in full. The principles used in building this photometer were employed by Pardue and Deming (646) in the modification of a Bausch & Lomb Spectronic 20 colorimeter for high stability, as described earlier in this section.

Chapter 4

DESIGN AND CONSTRUCTION
OF CONTINUOUS FLOW MANIFOLDS

The selection and assembly of the tubing and fittings comprising a continuous flow manifold demand the most creative and intuitive qualities of an analytical mind. Rarely does the first attempt at constructing a new manifold produce a satisfactory result, and diagnosing exactly where one went wrong often requires the closest observation of the characteristics of the flowing stream. The problem is compounded by the wide variety of fittings, and thus choices, available to the manifold designer. It is perhaps this very combination of choice, creation, and frustration that gives rise to the fascination and hopefully the ultimate satisfaction in obtaining a smoothly running continuous flow manifold.

Fortunately, much has been written on the topic of manifold design. We shall begin our review of the literature by considering a few general areas concerning the construction and evaluation of manifolds. Next we shall cover the details of manifold construction and, to avoid drowning in alphanumeric seas of A6s, B0s, and PC1s, we shall approach the subject by functions, progressing from the simpler operations, such as segmentation and debubbling, to the more complex, e.g., extraction and distillation. We shall then mention some characteristics of tubing and specialized fittings and shall conclude the chapter by describing several interesting manifold designs.

119

I. GENERAL CONSIDERATIONS

A. Temperature

Temperature control of continuous flow systems may be desirable or necessary for several reasons. One may be dealing with samples or reagents which decompose too rapidly at room temperature. Jocelyn (438) placed a Sampler, Proportioning Pump, reagents, and manifold in a refrigerator to hold the temperature at about $4^{\circ}C$; the Dialyzer, Colorimeter, and Recorder were operated outside the refrigerator.

The timing of peaks may be critical, especially when timers must be used to synchronize associated devices with the operation of the Auto-Analyzer. Pollard and Waldron (666) placed delay coils in a constant temperature bath to avoid changes in the volume of air bubbles and thus to improve timing accuracy in the flowing stream.

Manifold leakage due to viscous reagents may be alleviated by adjusting the flow stream temperature. Hochella and Weinhouse (404) used a reagent containing concentrated sulfuric acid which was chilled by being passed through a jacketed mixing coil. The cold water used in the coil jacket was warmed slightly, by first passing it through a 10-ft mixer coil and then feeding it to the coil jacket to reduce the viscosity of the acidic reagent slightly and thus to avoid leaks.

Finally, temperature control becomes a necessity if one or more of the operations on the manifold do not go to completion and if the degree of completion is a sizable function of temperature. Hardman (378) operated an AutoAnalyzer in a mobile van and correlated sinusoidal drifts on the Recorders with the variation in outside temperature actuating the van's internal air conditioner. The problem was corrected by placing transparent Lucite covers over the Proportioning Pumps and manifolds to hold their temperature constant. Wade and Phillips (879) found that the temperature of the manifold coils rose when the Proportioning Pump motor was turned on, affecting the analytical results. They provided a degree of insulation by mounting the critical manifold components on supports made of 1-in.-thick expanded polystyrene. Hales et al. (365) placed mixing coils in a constant temperature bath to stabilize the amount of carbon dioxide dissolved from a gas phase into a liquid colorimetric reagent. Kendal (465) recommended immersion of glass pulse suppressors in constant temperature baths. Von Berlepsch (874) combined a single mixing coil and an H0 fitting into one water-jacketed piece.

B. Light

Light may also adversely affect the chemistry of a manifold. Berry and Crossland (78) shaded mixing coils from direct sunlight to avoid decomposition of diazonium salt solutions. Bano and Crossland (58) wrapped an extraction coil with black paper to prevent photodecomposition of dithizone in chloroform. Fleet et al. (281) placed black cloth around coils, transmission tubing, and reagent reservoir used for a hydrogen peroxide solution. Rokos et al. (688) placed critical mixing coils inside a box to exclude light which otherwise decreased the fluorescence signal in an AutoAnalyzer method for serum creatinine phosphokinase activity.

C. Evaluation of Manifolds

In evaluating the characteristics of a manifold, one may find that other components of a sample matrix interfere in the analysis for the material of interest. If the degree of interference is low enough, one may consider using a "leveling" technique. For example, Wahl and Auger (881) noted that fluoride and carbonate interfered in the determination of sodium hydroxide. The problem was solved by adding sodium fluoride and sodium carbonate to one of the manifold reagents in amounts far exceeding those present in the samples. The continuous pumping of the reagent to the manifold produced a reproducible maximum interference which was automatically accounted for when the sodium hydroxide standards were processed through the manifold.

In a somewhat related technique, the recovery of an analyte in the presence of the sample matrix may be evaluated by continuously aspirating a solution of the analyte through a diluent line (767). After the raised baseline stabilizes, the samples and standards are run through the system, resulting in responses on top of the raised baseline, and the sample values are calculated. The normal diluent (i.e., without added analyte) is then pumped to the manifold. After the baseline returns to normal position, the samples and standards are again processed, and the sample results are again calculated. The two sets of results, with and without the added analyte, are compared to estimate the amount recovered. Hill and Kessler (396) used this technique to estimate the recovery of glucose. A series of samples was analyzed normally and then reanalyzed while a solution of glucose was continuously added through a diluent line. Equivalent results with and without the added glucose were interpreted as 100% recovery. Gitelman et al. (329, 330) evaluated the recovery of magnesium from biological samples in a similar fashion. This procedure, which may seem a bit odd when first considered, is simply the "continuous" version of a well-established technique, that of standard addition.

Another useful bit of information in evaluating a newly designed manifold concerns whether a color reaction has gone to completion or is stable. Wrightman and Holl (921) describe a convenient procedure. The system is allowed to reach steady state by continuously aspirating a standard solution. The tube carrying the color solution to the flowcell is disconnected, but the recorder is left in operation. Any change in recorded absorbance quickly yields the desired information. This test may be repeated with several samples to determine whether the sample matrix changes the result.

Laessig et al. (514) studied the Technicon SMA 12/30, SMA 12/60, and Hycel Mark X analytical systems. These instruments all utilize the "two set point" calibration method; that is, they are calibrated at zero and at the value of one standard material. Thus, the operator must be able to rely on the linearity of the instrument. Since none of these instruments is perfect, it becomes of interest to determine the level of the standard that will result in the best overall accuracy. In a perfectly calibrated analytical system, the intercept of the standard curve should pass through zero, and its slope should reflect a 1 to 1 correspondence between the standard strength and the instrument response. Neither ideal can always be attained, but one may approach them closely. Laessig et al. (514) give the results of collaborative studies on the determination of bilirubin and discuss a method for determining the best "set point" using a control serum and the standard addition technique.

The reduction of carryover (sample interaction) to a low value is almost always a major goal in designing a continuous flow manifold. We shall touch on the practical aspects of this subject at several points in this chapter. For the moment, we note that residual carryover may be dealt with by graphical or mathematical corrections. An example of the former is given by Kemp (464) and of the latter by Wallace (888). Broughton et al. (138) recommend a method for evaluating automatic analyzers, including tests for sample interaction. The theoretical aspects of sample interaction are discussed in Chapter 7.

II. MANIFOLD FUNCTIONS

A. Sampling

Scholes and Thulbourne (733) describe a method of sampling wherein rather large dilutions may easily be made. In their example, the diluent is pumped at 0.32 ml/min. From the exit of the pump tube the diluent is led to a T fitting attached to the inlet of the sample pump tube. The third arm of the T fitting is connected to the sample source line. The sample pump tube pumps sample plus diluent at 0.42 ml/min; thus, the actual sample flow rate is 0.10 ml/min. As the sample and diluent leave the sample pump tube, air bubbles and further diluent (or a reagent) are added, and the whole is passed through a mixing coil. A double dilution is thereby effected without the use of debubbling and resampling. Similarly, Lingeman and Musser (541) pumped isopropanol at 0.4 ml/min to a T fitting at the inlet of a sample pump tube rated at 1.1 ml/min; the effective sampling rate through the third arm of the T was 0.7 ml/min. The isopropanol helped wash out the sample pump tube and improved sample separation. Platt et al. (664) describe a variation of this scheme wherein the diluent is added at the Sampler probe itself. A T fitting is used in place of the usual probe. The bottom arm of the T dips into the sample, the diluent is added at the side arm of the T, and both sample and diluent are pumped out of the upper arm of the T. For example, Platt et al. pumped 2.9 ml/min of diluent into the side arm and 3.9 ml/min of diluent plus sample out of the upper arm. This system of sampling was felt to give faster rise times and less trailing off after peaks.

Bernhard and Macchi (75) operated an AutoAnalyzer on board a ship to analyze ocean water for phosphate, nitrite, and nitrate. To obtain samples through a 100-m hose, they designed a fitting made from Tygon and PVC tubing which segmented the water with air bubbles at the point of sampling.

Occasionally, samples must be obtained from highly hostile environments. Baumann and Roberts (67) used an AutoAnalyzer to measure the concentration of free sulfuric acid in a wet process phosphoric acid digestor. They describe a special continuous sampling head, suitable for use in a hot, acidic, supersaturated solution while monitoring on-stream chemical plant processes.

Removal of particulate matter from the sample stream may be accomplished by decantation. This operation is mentioned in Chapter 2, Sec. XII, A. Whitehead et al. (907) analyzed for ammoniacal nitrogen in river water. Their article shows a figure of a solids separator, a 0.25-in.-i.d. glass T fitting, which eliminated most of the suspended solids in the sample stream before the stream was led to the manifold.

B. Segmentation

Not all manifolds require air bubble segmentation. Manifolds which run continuously at steady state, such as those monitoring factory stream processes (4) or chromatographic columns (223, 441, 442), have been designed to operate without air bubble segmentation. Adelman and Pellisier (4) segmented the incoming sample stream with air and then immediately debubbled the stream to remove any dissolved air. The unsegmented stream then entered the manifold. The occasional bubbles present in the sample stream would otherwise have disturbed the flow of unsegmented liquids on the manifold. Other examples of unsegmented manifolds are described by Blaedel and Hicks (102, 103) and Bomstein et al. (122). Several authors have reported that segmentation by a second immiscible liquid phase (e.g., in extraction processes) substitutes adequately for air bubble segmentation (58, 99, 505, 507, 612, 613, 887).

1. Purpose of Segmentation

The primary reason for air bubble segmentation is to reduce carryover (sample interaction). Walker et al. (884) describe the axial flow of a liquid through tubing. The linear velocity is greatest at the center of the stream, the portion near the tubing wall being retarded by friction. The difference in linear velocity causes mixing and carryover. Although air bubbles greatly reduce this mixing effect, the thin film of liquid between the bubbles and the tube wall still contributes to carryover. However, most of the carryover occurs in the unsegmented portions of the manifold flow stream, and such portions should be shortened or eliminated whenever possible.

2. Air Scrubbers and Alternative Gases

The vast majority of published manifold designs incorporate air bubble segmentation, and with but few exceptions the air is pumped directly from the laboratory atmosphere. The AutoAnalyzer thus breathes the same air as the laboratory personnel. Opening a bottle of a volatile chemical (e.g., ammonia or concentrated hydrochloric acid) near an AutoAnalyzer may lead to the loss of several sample results if the manifold chemistry is at all affected by such reagents. In such cases, it is good preventative practice to scrub air free of reactive gases before using it for segmentation. The most common method is to allow the air pump tube to draw its air supply as bubbles through an appropriate reagent placed in a scrubber bottle.* Dilute sulfuric acid (from 1 to 3.5 $\underline{N}$) has been used for removal of ammonia, and aqueous sodium or potassium hydroxide (about 1 $\underline{N}$) has been used for removal of carbon dioxide. Alternatively, tubes containing ascarite and calcium sulfate (313) or Mallcosorb (Mallinckrodt) (468) may be attached to the air pump tube inlet to remove carbon dioxide. A tube containing a "molecular sieve" has been so used to remove carbon monoxide (800).

In manifolds designed to monitor pollutants in the atmosphere, a fitting is often used to absorb the desired analytes from the air sample into a liquid reagent. If the absorber fitting removes the analytes quantitatively, the excess purified air may then be used to segment the appropriate manifolds (928).

Gases other than air have been used for segmentation bubbles. Blackwell (100) segmented a stream with nitrogen to remove oxygen which otherwise depressed the colorimetric response of the system. Lenard

---

*For examples of manifolds using washed air for segmentation of manifold flow streams, see Refs. 6, 54, 55, 94, 158, 197, 202, 292, 293, 342, 352, 493, 546, 551, 573, 674, 768, 774, 828, 849, 881, 903, 915, and 919.

et al. (528) used nitrogen bubbles to reduce deterioration of ninhydrin reagent by air oxidation. The nitrogen was bubbled through the ninhydrin reagent bottle before being led to the nitrogen pump tube. Brandon (129) describes how to aspirate carbon dioxide for bubble segmentation from a pressurized gas cylinder.

3. Segmentation Parameters

There is no easy rule of thumb to guide us in choosing the ratio of air to liquid, for too much as well as too little air may degrade manifold performance. Jordan (443, 444) found that increasing the rate of air addition gave better wash, i.e., less sample spreading and better sample separation. Varley (869) noted that reducing the amount of air in mixing coils resulted in a better signal to noise ratio, that is, less noise on the recorded baseline. Gardanier and Spooner (306) studied the effect of air bubbles on longitudinal mixing in various AutoAnalyzer circuit components. They compared the wash characteristics of a Dialyzer, 3 yd 5 in. of tubing (0.0625 in. i.d., 5.3 ml volume), a double mixing coil (5.3 ml volume), and a Heating Bath coil (24.2 ml volume). The Dialyzer showed the best wash characteristics. The best wash in the Heating Bath coil resulted at high air bubble rates. In general, the wash became poorer when the air to liquid ratio dropped below 1:10. At high air flow rates, the double mixing coil, tubing, and Heating Bath coil all had the same wash characteristics and the Dialyzer was still better than the others. Walker et al. (884) varied the ratio of air to liquid and found that the manifold wash character- istics deteriorated when the ratio exceeded 1:1, i.e., when more air than liquid was used. Further comments on the effects of the air to liquid ratio may be found in Chapter 7.

The presence of air bubble segmentation raises the pressure required to achieve a given linear velocity through the manifold tubing. Chaney (177) has written a most instructive article on this phenomenon. Forty-foot lengths of tubing made of various materials were compared in tests which measured pressure drop versus linear flow velocity of water. Without air bubbles, all the tubing materials gave rather similar responses; that is, the pressure required to achieve a given flow rate did not differ much from one to the next. With air bubbles present, higher pressures were needed to obtain a given linear velocity, and marked differences appeared among the various tubing materials. Glass tubing required the lowest pressure, Teflon the highest, to achieve a given velocity. The addition of a compound which lowered the water surface tension (Brij) reduced the pressure requirements. These observations led Chaney to conclude that surface tension and the contact angle between the liquid and the tubing surface were the principal causes of the increase in pressure drop needed to maintain a given linear velocity when air bubbles were present. The force needed to push a segment along through the tubing is transmitted through the bubbles, and in the process their shape is distorted away from their equilibrium shape, i.e., the shape determined by surface tension and contact angle between fluid and tubing surface when the stream is at rest. Glass, having a low contact angle, allows the fastest linear flow of segmented stream at a given pressure, but suffers from a comparatively large transfer rate between segments (carryover) due to the ease with which it is wetted. Teflon and polyethylene are not wetted, but they build higher backpressures. Air bubble patterns are also harder to maintain in these latter materials, especially at higher velocities.

Chaney (177) concluded that conventional manifolds perhaps use too many bubbles (about 50 to 100 segments per sample) and recommended reducing the bubbles from the usual range (6 to 12 per foot of tubing) to 1 per foot, thereby reducing the backpressure developed in the manifold. The diluting effect of transfers between segments (arising from the liquid film between the bubbles and the tubing wall) was felt to be complete after 4 to 5 segments were formed; thus, further segmentation would not appreciably help to reduce that portion of carryover due to transfers between segments. In a mathematical model of flowing segment cross-contamination in AutoAnalyzer manifolds, Begg (69, 70) calculated that the degree of dilution of the last segment is almost independent of the number of segments into which the stream is divided, as long as at least 6 segments are present. This result is in good agreement with Chaney's observations.

A few points should be added to the above discussion. First, it is not always easy to reduce the rate of bubble addition to only five or so per sample. Second, once a manifold is in operation, bubbles are being added at a constant rate; therefore, as the Sampler rate is increased, the number of bubbles per sample is decreased. As the sampling rate is increased, carryover increases. At the point where carryover becomes too large to be tolerated, one should check the number of segments being formed. At very high sampling rates, it is possible that less than five or so segments will be formed per sample. Perhaps an increase in the air bubble rate will help. Third, the backpressure and cushioning action of air bubbles often act as an effective form of pulse suppression; thus, a reduction in the number of air bubbles may result in an increase in pulsing noise observed on the recorded trace. Lorch and Gey (551) recommended relatively high air volumes in the bubbles to achieve lower pulsation, better mixing, and more uniform sample distribution in the liquid segments. Finally, compounds which increase surface tension may increase the leakage between segments in a manifold system (399).

4. Effect of Tubing on Bubble Patterns

As mentioned briefly above, the tubing material selected may be a factor in the stability of air bubble patterns flowing through a manifold. Thiers et al. (831) reported that bubble patterns were far more stable in glass than in Tygon tubing. Pyrex transmission tubing (4 mm o.d., 2 mm i.d.) was used wherever possible on the manifold. Amador (22) found that polyethylene can be substituted for glass, retaining excellent bubble patterns, whereas Kel-F and Teflon have different wetting properties and cause breakup of bubble patterns. Levy and Keyloun (536) used polyethylene to eliminate air bubble pattern breakup which occurred in glass. Undoubtedly the properties of the liquid stream (aqueous versus organic, acidic versus alkaline, etc.) play a great part in the formation and maintenance of stable bubble patterns; often, trying a different tubing material may solve a stubborn case of bubble pattern breakup.

5. Special Fittings

Several authors have described special fittings for the injection of air bubbles into a flowing stream. Shaw and Duncombe (758) give a picture of a T fitting used to obtain regular air bubble patterns. Gerke et al. (324) enlarged the opening of a "conventional bubbler" to 4 mm to produce large air bubbles that would not break up as the stream flowed through Dialyzer modules.

Lane and Mavrides (517) obtained consistent bubble patterns by inserting thin polyethylene tubing about 1 in. into the bore of an A0 fitting and feeding pulse-suppressed air through the tubing. This arrangement gave small bubbles and small liquid segments. Amador et al. (23) show drawings of a "bubble regulator" made from a short length of polyethylene tubing inserted into an N5 nipple. The nipple was placed in a D1 or H3 fitting. The polyethylene tube extended into the glass fitting and injected bubbles of the desired size into the stream. The free end of the polyethylene tube was adjusted to extend just past the side arm entry of the glass fitting.

Neeley et al. (623) give a drawing of a "bubble-producing" device made of a plastic block equipped with a plastic nipple for the liquid entrance and a stainless steel tube (0.81 mm i.d., 1.27 mm o.d.) for the air entrance. The liquid enters the fitting and makes a 90-deg bend, placing it on the same axis as the stainless steel tube. The air bubbles are thus injected coaxially into the liquid stream. The device produced a uniform pattern of small air bubbles.

Pollard and Waldron (666) employed very slow flow rates and noted that the use of miniature bubble-forming fittings (PT4s) helped prevent mixing and carryover at the air-liquid junctions.

Rush et al. (702) used an alkaline, organic flow stream (0.8 M aqueous potassium hydroxide and 80% aqueous isopropanol) and found that bubble-forming fittings made of acrylic plastic were attacked by the solvent system. They show a drawing of an "air gun" device which injected the bubbles into the center of the flow stream and which was compatible with such solvent systems. The device was used with the air bar on a Proportioning Pump III.

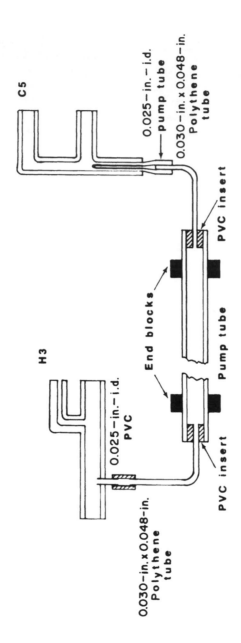

FIG. 1. An example of good technique when debubbling and resampling on AutoAnalyzer I manifolds. The volume of the C5 debubbler is minimized; narrow-bore tubing is used on the unsegmented stream; the pump tube carrying the unsegmented stream is trimmed close to the end blocks; the fittings are placed close to the pump tube inlet and outlet. The solution flows from right to left. Reproduced from Sawyer and Dixon (724) with the permission of the authors and the Society for Analytical Chemistry.

6. Suspensions

Segmentation of a reagent stream may be useful in handling suspensions. Gales et al. (299) kept a slurried reagent mixed by a magnetic stirring bar arrangement. As the reagent was pumped from its supply vessel, it was immediately segmented with air to keep the reagent moving uniformly.

C. Debubbling and Degassing

Debubbling and resampling of flow streams should be avoided if at all possible, because the mixing action in the debubbler chamber contributes heavily to carryover (884). Whenever this operation must be used, one should place the debubbler near the inlet of the next pump tube to minimize the length of the unsegmented stream from the debubbler (518). Christopherson (185) used small-bore Teflon tubing (thin-wall, 24-gauge) on all unsegmented streams to improve sample resolution and described how to connect the Teflon tubes to standard glass fittings via nipples or sleeve connectors. An analogous situation occurs at the debubbler just ahead of a flowcell, and the reader is referred to the discussion in Chapter 2, Sec. II, B, 4, b.

1. Modified Debubblers

Jordan (443, 444) modified debubbler fittings to reduce their mixing action. The inlet chamber size was reduced to the minimum value which would still permit proper debubbling. The debubbled liquid was withdrawn through capillary tubing to move the stream rapidly to the next point on the manifold. Sawyer and Dixon (724) show a drawing of a modified C5 debubbler (Fig. 1). Narrow-bore polythene tubing (0.03 in. x 0.048 in.) was inserted into the lower limb of the C5 and attached to it with a short piece of 0.025-in.-i.d. pump tubing. The narrow-bore tubing led the debubbled stream to the next pump tube. A similar piece of narrow-bore tubing led the unsegmented stream from its pump tube to the next fitting on the manifold. The purpose again was to minimize the length of the unsegmented stream. Similarly, Ward and Hirst (890) give a figure of a low dead volume debubbler made from a standard A0 fitting by inserting very thin polyethylene tubing into the A0 leadoff arm. The polyethylene tubing was held in place by a short piece of Tygon tubing acting as a sleeve.

A few specialized separation fittings have been reported. Duncombe and Shaw (234) show a miniature glass "cyclone" used to separate liquid and gas phases under conditions of high gas flow rates. These authors also have published a picture of a vacuum jacketed gas-liquid separator (758, 759). Other gas-liquid phase separators are mentioned below in Sec. II, K.

2. Degassing

There are two general procedures for degassing liquid streams in continuous flow manifolds. First, one may pass the stream through a Heating Bath to release dissolved gases (447, 448). Second, one may add air bubbles and then debubble the stream (4). Feller et al. (259) used chemical reactions which released gas bubbles at one point on the manifold. By adding a double delay coil between the manifold and the Colorimeter, enough time was provided for the air segmentation bubbles to absorb the extra gas bubbles. Knowles and Hodgkinson (493) reported a method for serum oxalic acid in which enzymatic decarboxylation released free carbon dioxide. The carbon dioxide was then separated and determined colorimetrically. To avoid the error introduced by the carbon dioxide already present in the samples, a continuous automatic degassing operation was carried out before treatment of the samples with the enzyme. An 8-m length of polythene tubing (1.77 mm i.d.) was coiled vertically around a 25-mm core support. The sample at 3.75 ml/min plus a stream of air free of carbon dioxide was fed into the top of the coil. The mixture passed downward through the coil in about 15 sec; the liquid phase was recovered through a C4 fitting and was then passed to the enzymatic reaction manifold. This system was operated at 15 samples per hour.

D. Mixing

Achieving a closer match between the densities of the two liquid streams to be combined may aid in more efficient mixing. For example, in developing an AutoAnalyzer procedure, Sankuer et al. (715) first tried adding a reagent (1, 2-dimethoxybenzene) dissolved in absolute alcohol to a sample stream containing a high concentration of sulfuric acid. Poor mixing resulted, as evidenced by an intolerable level of noise on the recorded trace. The noise level was reduced somewhat by selecting a different solvent (isoamyl alcohol) for the reagent; the density of this solvent more closely approached that of the acidic sample stream.

Lane and Mavrides (517) reduced sample interaction by using large ratios of reagent volumes to sample volumes.

1. Arrangements of Fittings for Stream Addition

Kenny and Jamieson (467) designed an AutoAnalyzer manifold which required mixing acetic anhydride with serum samples. Enough heat was generated at the mixing point to attack the usual flexible connections. The problem was solved by fusing the glass fittings directly together to give an all-glass system. A photograph of the arrangement appears in their paper.

Efficient, rapid mixing of liquid streams is often aided by careful selection of the fitting used to introduce one stream into the other. The insertion of a small length of tubing into a glass junction fitting has been found useful by several authors. Hill (394) used a 0.020-in.-i.d. tube inserted into the base of a T fitting to introduce a slowly flowing sample stream (0.16 ml/min) into a rapidly flowing reagent stream (11.7 ml/min). The sample tube was adjusted so as to project into the reagent flow stream. Morgenstern et al. (608) inserted polyethylene tubing (0.011 in.) into an H0 fitting to introduce a sample stream into a substrate stream. Thomas (837) injected a sample stream into a reagent stream by fitting a short length of Teflon tubing (0.01 in. i.d.) into the outlet of the sample pump tube, leading the Teflon tubing into a D1 fitting. This arrangement improved the uniformity of mixing. Young (925) mentions adding a sample stream to a diluent stream through an H1 cactus provided with a fine polyethylene insert to ensure uniform distribution of the sample (serum) into the diluent (0.9% aqueous sodium chloride containing 10 drops of 10% Brij per liter).

Jones (441, 442) shows a special fitting used to mix streams. Four stainless steel tubes are imbedded in a 1-in. cube of plastic at right angles in a common plane. A very small chamber is left open at the center junction point, and the device thus has a very small dead volume.

Klein and Kaufman (485) found that the use of an A6 fitting promoted the smooth entrance of liquids at small flow rates into liquids of larger flow rates. This fitting features a short platinum capillary inserted into a 25-mm length of 2-mm tubing.

At one point in a manifold, Jenner (437) needed to mix a sample stream of relatively low density with a reagent stream of higher density. In the addition, any backpulsing of the reagent into the sample stream caused the reagent to precipitate. Jenner gives a figure of a special fitting which caused the sample stream to mix downward into the reagent stream. At the juncture, the reagent stream flowed through a capillary; its flow rate thus was increased but its pressure was decreased. This lowering of pressure prevented backpulsing of the reagent into the sample and corrected the precipitation problem.

Hill and Cowart (395) described a method in which two reagents, kept chilled in their reservoirs, required mixing on the manifold just prior to the addition of the sample stream. After the reagents were mixed, but before the sample was added, the combined reagent was led through an "overflow sampler" fitting, allowing the mixture to equilibrate with atmospheric pressure. An unstable baseline resulted if this equilibration was not carried out. A picture of the overflow sampler appears in their paper.

The addition of immiscible liquid phases is discussed below in Sec. II, G, 1.

2. Mixing Techniques

Having combined the liquid streams, one generally completes the mixing action by passing the stream through one or more mixing coils. White et al. (674) point out that mixing coils function properly only when used in a horizontal position. Varley (869) studied the relationship between the ratio of liquid to air and the mixing efficiency in mixing coils and noted that reducing the air resulted in a better signal to noise ratio (less noise on the recorded baseline). The optimum air volume was found to be 12.5% that of the liquid. The air was distributed in such a fashion that there was at least one bubble in each half-section of the coil. Less than one air bubble in each half-section would greatly reduce mixing efficiency. When the amount of air was kept constant at 12.5%, more noise resulted from using a wide-bore tube to introduce the air than from using a capillary tube. Beaded coils have been used for a final mixing action just ahead of the Colorimeter (828, 837). A 40-ft delay coil laid on its side has been substituted for several mixing coils in series (556, 558).

In adding 75% sulfuric acid to a flow stream, Cadmus and Strandberg (159) found that feeding the air-segmented stream directly into the acid stream caused noise on the recording, probably due to striations in the Colorimeter flowcell. At first, a C1 debubbler was used to resample the stream before the acid stream was added, and then the combined stream was resegmented with air. Mixing the unsegmented streams solved the striation problem. However, a standard C1 debubbler degraded the wash characteristics of the system. The wash was partially restored by inserting a 0.030-in.-i.d. tube into the C1 at the point of resampling, thus reducing the internal volume of the fitting. A figure of this arrangement is given in Cadmus' and Strandberg's article.

Kurtzman et al. (502) achieved a rapid (10-sec) agitation of a sample stream with a reagent stream (4.62 ml/min total) by adding air at 11.7 ml/min, passing the stream through standard mixing coils and then immediately debubbling the stream before the next operation on the manifold.

3. Special Mixers

A useful mixing device consists of a fitting containing alternating capillaries and chambers, the so-called jet mixer. Jet mixers may be constructed from glass or plastic tubing. In the latter, short lengths are cut from two pieces of tubing of different internal diameters and press-fitted together to form alternating confined and expanded segments (525, 906). Typical combinations of tubing are 0.022 to 0.080 in. i.d. (775), 0.035 to 0.081 in. i.d. (662), and 0.035 to 0.090 in. i.d. (358). The last-mentioned combination was recommended for relatively high flow rates. Glass jet mixers have also been reported in several papers (234, 758, 821, 852). They may be made from glass transmission tubing by softening the glass in a flame to form fairly regularly spaced capillary sections (312). Jet mixers may be substituted for mixing coils or may be used along with mixing coils to handle difficult liquids. They are generally, but not always, used before the addition of air segmentation bubbles.

Morgan et al. (606) described an all-glass conically shaped mixer which proved useful in mixing an aqueous stream with a reagent containing about 80% alcohol. The streams were combined through a T fitting; they then entered the top of the conical mixer and were satisfactorily mixed upon exiting at the bottom. These authors also used broken glass beads or Teflon shavings inside lengths of 3-mm glass tubing for mixing aqueous and nonaqueous streams.

Magnetic stirring systems have been used in continuous flow manifolds (135). Skinner and Docherty (776, 777) describe a small chamber containing a magnetic stirring bar which was used to give a stream a final mix before being sent to the flowcell. The device includes an integral debubbler. Skeggs (773) mentions the use of small in-line magnetic mixing fleas for mixing liquid streams too stubborn for conventional mixing coils or jet mixers. Schuel and Schuel (741) activated magnetic fleas within the lines with a solenoid to induce thorough mixing.

Blaedel and Hicks (103) described a totally different approach to mixing. In their manifold (which used no air segmentation), a reagent flowed by gravity to a T where it met the sample stream pushed to the T by a peristaltic pump. To achieve good mixing, the tubing carrying the reagent was acted on by a pulser made of a nylon disk mounted slightly off center on a motor shaft turning at 300 rpm. As it rotated eccentrically, the disk depressed and released the gravity flow tubing, thus causing periodic reversals in the direction of flow. The improved mixing gave a faster approach to steady state.

E.   Stream Splitting

Several interesting fittings have been described for stream splitting. Nishi and Rhodes (628) show a figure of an acrylic block (0.5 in. x 1 in. x 1.5 in.) containing two channels at right angles in a common plane. One channel was 0.0550 in. wide; the other was 0.0292 in. wide. The stream enters the wider channel, and some flows straight through to waste. Portions of the stream are drawn off through the exits of the narrower channel to each of two manifolds. Countersunk nipples were provided for tubing connections.

Wrightman (919) gives drawings of two devices, each made from an A4 fitting; one device yields one stream (a resample fitting), and the other

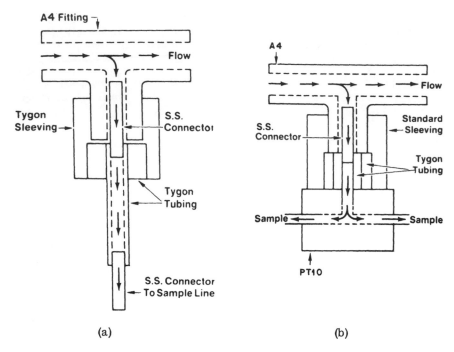

(a)   (b)

FIG. 2. Wrightman's stream splitters. S. S. Connector stands for N13 stainless steel connector. (a) One stream resampled. The upper N13 connector is adjusted so that its end is about 0.0625 in. below the horizontal wall of the A4. The A4 is mounted so that its side arm extends vertically downward. (b) Two streams resampled. A PT10 is butted directly to the N13 connector extending downward from the A4. In both (a) and (b) the incoming air bubbles exit at the upper right of the A4. The resampled streams are unsegmented. The low internal volumes of the fittings helped to maintain sample integrity while debubbling and resampling. Reproduced from Wrightman (919) by permission of the Technicon Corporation.

yields two resampled streams (Fig. 2). These devices were found to give highly efficient resampling actions, thus maintaining good sample separa- tion. Hoyt and Jordan (416, 417) show a figure of a combination debubbler/ stream splitter fitting made by adding a capillary T to the lower portion of a C5 debubbler. The fitting minimized dead volume, both in debubbling and in stream splitting, thus maintaining good wash.

Kawerau (457) described how to construct an in-line stream splitter from standard Technicon tubing and fittings. Capillary tubing was inserted inside of normal transmission tubing so as to remove liquid but allow air

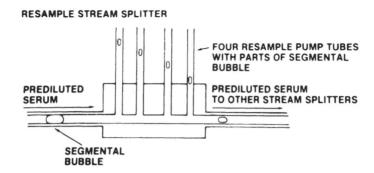

RESAMPLE STREAM SPLITTER

FOUR RESAMPLE PUMP TUBES WITH PARTS OF SEGMENTAL BUBBLE

PREDILUTED SERUM

PREDILUTED SERUM TO OTHER STREAM SPLITTERS

SEGMENTAL BUBBLE

FIG. 3. A five-way stream splitter which maintains segmentation on all five resampled streams. Reproduced from Crerand et al. (203) by permission of the authors and the Technicon Corporation.

bubbles to pass by. This arrangement allowed stream splitting to various manifolds without using the ordinary debubbler fittings. In the example given, two capillaries were positioned inside the transmission tubing, one "upstream," the other "downstream" (nearly touching), to sample the flowing stream at precisely the same point. In a later paper, Kawerau (456) shows a photograph of a stream splitter in which two capillary tubes are inserted side by side into a short piece of tubing, allowing withdrawal of two sample streams from the same segment of sample.

Crerand et al. (203) give a drawing of a stream splitter which operates on an incoming segmented stream and delivers five output streams, each containing a portion of each incoming bubble (Fig. 3).

F. Pulse Suppression

Pulsing occurs as the leading pump roller releases the pump tubes. The backpressure exerted by the manifold momentarily pushes liquid or air back into each pump tube. Technicon supplies a very effective pulse suppressor fitting for air lines.

Narrow-bore tubing may be inserted in the manifold flow stream to reduce pulsing (674). Generally this tubing is placed between the pump tube exit and the next fitting on the manifold (83, 86). For air line pulse suppression, 0.005-in.-i.d. tubing has been used in lengths from 1.5 to 7.5 in. (123, 367, 517). Levine and Zak (533) connected a standard 0.005-in.-i.d. pump tube between an air pump tube and the next glass fitting to act as a pulse suppressor. Larger tubing is used for liquid lines,

0.015 in. i.d., for example (123, 367). Sparagana et al. (785) used the following combinations of tubing for liquid line pulse suppression:

Pump tube (in. i.d.)	Pulse suppressor tube (in. i.d.)
0.065	0.045
0.073	0.040
0.073, 0.090, 0.110	0.056

Moorehead and Sasse (603, 719) placed a pulse suppressor between a Heating Bath and a Colorimeter. Nipple connectors were used to insert a 4-cm length of 0.25-mm-i.d. (0.01-in.-i.d.) manifold tubing between two lengths of transmission tubing.

Kawerau (456) shows a photograph of a pulse suppressor made from a length of narrow-bore tubing, doubly coiled and held neatly by being threaded through a sample cup cap in which a hole had been punched. The coils flexed with the action of the Pump and absorbed the pulse waves. McCullough et al. (584) give figures of two pulse suppressors, used for air or liquid lines, made from variously sized tubing and featuring pinholes punched on the inlet sides. These pulse suppressors gave very smooth flows of reagents and air.

Fittings similar to the jet mixers described above in Sec. II, D, 3 have also been used as pulse suppressors. Lorch and Gey (551) show such a device, used on liquid lines, made from tubing of various diameters and stainless steel nipple fittings. Hinton and Norris (399) used a glass pulse suppressor on a liquid line. It was a 6-cm length of 3-mm-i.d. tubing constricted to about 0.1 mm i.d. in four places.

Christopherson (185) made constrictions in the exit arms of A1 and D2 fittings to act as pulse suppressors and micro mixing chambers. Adelman (3) mentions the use of a quarter-length delay coil to reduce pulsing and obtain smoother recordings. The coil was placed on its side and the stream flowed through the coil just before entering the Colorimeter. Shaw and Duncombe (758) give a photograph of a tapered mixing coil used in the direction of increasing coil diameter to suppress surging from the Pump.

Two mechanical flow regulators have been reported for use with continuous flow manifolds. Valentini (856) shows a figure of a Teflon "micropump" containing a Teflon membrane and a plunger activated by air pressure. From the Pump, the liquid passes by the membrane and its

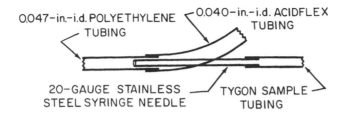

FIG. 4. A fitting to inject a sample stream into an immiscible phase. The sample stream enters through the stainless steel needle. It is segmented into the extractant stream (in this case, ethylene dichloride), which enters through the Acidflex tube. The stream flow is from right to left. The fitting is based on a design by Kuzel (505, 507). Reproduced from Stevenson and Comer (795, 796) by permission of the authors, the Technicon Corporation, and the copyright holder (the American Pharmaceutical Association).

flow (pressure) may be regulated by the amount of air pressure exerted on the plunger. Isreeli (428) described a nonpulsing flow controller for continuous flow systems. The reagent is pressurized, in a large plastic bottle, for example. The reagent flows into the device and, after passing through a series of balanced diaphram valves and springs, emerges at a constant flow rate to the analytical system. The device is essentially a "hydraulic closed-loop feedback servo control" regulator. A drawing and photograph of it are given in Isreeli's paper. Flow rates of 0.1 to 10 ml/min could be obtained. Long-term flow control (weeks to months) was ± 0.5 to $\pm 1\%$. Short-term variations were too low to be measured. The response of the device to changes in inlet pressure was fast enough that changes in the analytical system could not be detected. Responses to changes in pressure at the outlet were between fractions of a second to a few seconds, depending on the flow settings.

Another alternative approach to pulse suppression on air lines consists of feeding pressurized air from a compressed air tank into the Pump's air tube inlets. This technique is discussed in Chapter 2, Sec. IX,A.

G. Extraction and Separation of Immiscible Liquid Phases

1. Injection of Immiscible Phase

Although standard glass fittings are generally employed to combine immiscible liquids, special fittings are sometimes useful. Stevenson and Comer (795, 796) show a drawing of a fitting designed to introduce an aqueous sample stream in segments into a stream of ethylene dichloride (Fig. 4). A stainless steel needle was used to inject the sample stream into the tubing carrying the ethylene dichloride. This arrangement was based on a design by Kuzel (505, 507), whose paper should also be consulted. Bano and Crossland (58) injected a heptane phase as small droplets into an aqueous stream through an A6 fitting.

2. Extractions in Coils: Parameters and Techniques

Sodergren (783) studied the effect of varying the size of chloroform segments in extraction coils. Pulsation in the coils was caused by the difference in specific gravity between the chloroform and aqueous segments and occurred when the chloroform segments circulating in the coils were too large. Such pulsation led to an irregular baseline on the recording. The pulsation was controlled by altering the size of the aeration tube so as to divide the chloroform into smaller segments.

Extraction of an aqueous phase with chloroform may also be carried out without air bubble segmentation, and Wallace (887) has written an interesting discussion of such an extraction system. The chloroform segments are forced into elongated shapes and do not contact the glass coil tubing, but instead slide along the thin film of aqueous phase which wets the tubing wall. This thin layer promotes rapid transfer of the extracted species between the liquid phases. The rapid transfer between phases acts in concert with the convection currents in each segment to produce rapid equilibration. This model predicts that the extraction would approximately follow first-order kinetics, the tubing length corresponding to time and the rate constant being a function of the ratio of the tailing rate to the volume of an aqueous segment. This extraction system was found to be much less prone to emulsions than systems using beaded coils. Wallace notes that air bubbles unfavorably influence the extraction process because they reduce the aqueous layering on the glass tubing and also disturb the segmentation pattern by causing the chloroform segments to queue up and merge. As noted in Sec. II, B, several other authors have found that segmentation by immiscible phases substitutes adequately for air bubble segmentation. However, Robertson et al. (687) mention that occasional air bubbles passing through the system may disturb the extraction equilibrium. In addition, other constituents of the sample matrix (e.g., excipients from drug samples) may upset the segmentation process. They therefore used air segmentation throughout the manifold, including the extraction coil.

As mentioned earlier, coils function properly as mixers only when used in a horizontal position. However, in the final stages of an extraction process, placing coils in a vertical position may aid in obtaining a smoothly flowing manifold. After dispersing steroid tablets in a SOLIDprep Sampler, Beyer (84) extracted the drug from the water slurry by adding chloroform and passing the stream through three solvent extraction coils. To obtain an even pumping action and to maintain satisfactory bubble patterns, the first two coils were placed horizontally and the final coil was positioned vertically. A small horizontal mixing coil was then placed between the vertical extraction coil and the phase separator fitting to promote satisfactory separation of the chloroform and water phases. Heistand (384) shows a photograph of an extractor coil, made from a beaded mixing coil, an H connector, and a pulse chamber, used in a vertical position. Petroleum samples plus concentrated sulfuric acid flowed down the coil, and nitrogenous materials were extracted from the petroleum into the acid. In an AutoAnalyzer method for sialic acids, Kendal (465, 466) added reagents to form a color and then extracted the colored species into butanol containing 10% by volume of 12 $\underline{N}$ hydrochloric acid. The extraction was carried out in a vertically mounted extraction coil, and Kendal gives a photograph of the coil plus a small volume phase separator, which yielded a clean recovery of the acidic butanol layer before it was passed to the Colorimeter. The presence of 10% by volume of 12 $\underline{N}$ hydrochloric acid in the butanol resulted in the formation of a solvent phase that separated well from the aqueous phase without forming refractive index gradients, thus helping to stabilize the recorded baseline (465).

Beaded coils have been used to promote contact between immiscible liquid phases in many AutoAnalyzer extraction procedures, including one of the earliest reports (654). Intimate contact between the immiscible liquid phases has also been achieved by placing glass wool plugs in both ends of a mixing coil (241). Ruzicka and Lamm (707) show a drawing of a glass fitting used in-line between mixing coils to improve contact between carbon tetrachloride and aqueous phases. The two phases flow into a small chamber and are forced out through a capillary hook which protrudes into the chamber.

Browett and Moss (139, 140, 610) designed a special glass extraction coil to extract lead from an aqueous solution into a solution of dithizone in chloroform and carbon tetrachloride. The coil was made from 10 ft of 2.7-mm-i.d. Pyrex tubing formed into a flat, horizontal 10-turn spiral.

Extractions (or reactions) between immiscible phases may sometimes be aided by passing the mixture through a heated coil. Fournier et al. (286, 287) combined an ethylene dichloride stream containing estrogens with a stream of diluted Kober reagent (1 part phenol in 4 parts concentrated sulfuric acid) and sent the mixed stream through a jacketed coil maintained at 70°C, using antifreeze as the heating fluid. Heating the coil provided three advantages: The transfer of the estrogens from the organic phase to the Kober reagent was promoted; the viscosity of the Kober reagent was decreased; and the subsequent separation of the phases was facilitated. Kream et al. (498) reacted a stream of steroids dissolved in ethylene dichloride with Porter-Silber reagent (phenylhydrazine solution) by passing the combined phases through water-jacketed glass coils held at 40°C.

The time required to reach extraction equilibrium may be shortened by placing the extraction coil(s) on a vibrator. Carter and Nickless (169) published an interesting study of the effects of coil vibration in a system using an extraction from an aqueous phase into a carbon tetrachloride phase. Data are given for the percentage of extraction versus shaking speed for four mixing coils connected in series. Extraction equilibrium (essentially complete extraction) was achieved above about 40 Hz, with a motion amplitude of about 1 cm. Without vibration, only 28% of the analyte was extracted, and it was calculated that about 50 stationary coils in series would have been needed to reach 98% of the theoretical recovery, an obviously impractical approach. Vibration speed was measured using a strobe light which also permitted observation of the contents of the coils. At 45 Hz, the carbon tetrachloride globules were vigorously agitated within their own section of aqueous phase, but the air bubbles still provided a fairly good barrier to diffusion. This observation was confirmed by allowing the system to reach steady state and then return to the baseline while fractions were collected at 30-sec intervals. The absorbance of each fraction was plotted, and a fairly sharp rise and fall profile was revealed, indicating that diffusion was minimal in the vibrating coil system. James and Townsend (432) carried out an extraction by passing two immiscible phases through plastic tubing that was agitated at 100 Hz by a special vibrator, a picture of which is shown in their paper.

Extraction coils made of materials other than glass have proven useful. Kuzel (505, 507) described a general AutoAnalyzer method for the determination of tertiary amines using a "dye complex" extraction. Kuzel evaluated about 10 dyes as to their retention or adsorption in AutoAnalyzer manifold glass and tubing components and found bromocresol purple (Eastman) most suitable. A Teflon extraction coil was used instead of a glass coil to reduce the adsorption of the dye. The coil was constructed by wrapping Teflon spaghetti tubing (0.038 in. i.d.)* around a support core. Kuzel found that the immiscible phases provided adequate stream segmentation and no air bubbles were needed in the extraction coil. Furthermore, the length of the coil could be varied (to find the optimum length) with very little effect on the amount of sample interaction. Because the mixing action was gentle, emulsions generally did not form, and a small volume phase separator could be used at the coil exit, further improving the system's wash characteristics. MacDonald et al. (560) used a similar Teflon extraction coil made from a 16-ft length of 0.038-in.-i.d. Teflon tubing wound on a cylinder of 1 in. diameter. Fournier et al. (286, 287)

---

*In one paper (505), the text gives 0.38 in. i.d. for the Teflon coil tubing, whereas the flow diagram gives 0.038 in. i.d. In another paper (507), 0.038 in. i.d. is given throughout. The correct value is 0.038 in. i.d. (private communication from Dr. Norbert R. Kuzel, April 10, 1974).

published an AutoAnalyzer method for the analysis of estrogens in the urine of pregnant mares. The urinary estrogens were extracted into an ethylene dichloride reagent stream and were subsequently determined colorimetrically. Glass extraction coils were found unsuitable because the urine adhered to the glass tubing, resulting in emulsions and an increase in the carryover of the system. A polyethylene extraction coil was substituted to overcome these problems. Kream et al. (498) extracted urinary steroids into ethylene dichloride using a Teflon mixing coil (1.6-mm-i.d. tubing, 21 turns on a 23-mm core).

3. Special Extraction Devices

Pinnegar (659) used an AutoAnalyzer to analyze beer samples. In an extraction step (acidic beer/isooctane) the presence of air led to the formation of stable emulsions. Pinnegar described a water-jacketed extractor device which provided intimate contact between the aqueous and organic phases in the absence of air. The extractor consisted of a glass tube, 10 cm in length and 3 mm i.d., containing about 15 coarse sintered polyethylene disks tightly fitted at equidistant intervals.

Ganis (303) shows photographs of small extraction vessels made from Erlenmeyer flasks with inlet and outlet tubes attached around the bottom and fitted with Teflon stirring bars to mix the immiscible phases. By using these extraction vessels, Ganis avoided the problem of clogging experienced with beaded glass extraction coils.

4. Separation of Immiscible Phases

Once the immiscible phases are mixed, the next operation on the manifold consists of recovering the desired phase through a separator fitting. As in the case of debubblers, the mixing action in the separator fitting can contribute significantly to the system carryover. The phase separator should be placed near the inlet of the flowcell (or of the next pump tube, etc.) and the length of the unsegmented stream should be minimized (706).

Obtaining a satisfactory separation of the desired immiscible phase from the exit of the mixer can be one of the most vexing problems in the design of a continuous flow extraction. Although the standard separator fittings often work well, several special separator systems have been reported to handle specific combinations of immiscible phases (see Table 1).

Arrangements which simultaneously separate the desired liquid phases and debubble the stream have been found useful. Stanton and McDonald (787, 788) used dithizone in benzene to extract metals from an aqueous solution and show a drawing of a separator fitting which debubbled the stream, recovered the benzene phase, and discarded the aqueous phase.

TABLE 1

Special Fittings for Separation of Immiscible Liquid Phases

Phases[a]	Reference
Amyl acetate; aqueous	Ganis et al. (304)
Aqueous; chloroform	Ellerington and Ferguson (241)
	Zagar et al. (929)[b]
Aqueous; chloroform:diethyl ether: pentanol (80:20:5)	Blackmore et al. (99)
Benzene; aqueous	Stanton and McDonald (787, 788)
Benzene:isopropyl alcohol (9:1); aqueous	Valentini (856, 857)
Butanol:12 $\underline{N}$ hydrochloric acid (10:1); aqueous	Kendal (465, 466)
Carbon tetrachloride; aqueous	Carter and Nickless (169)
	Sebborn (753)
Carbon tetrachloride, chloroform; aqueous	Browett and Moss (139, 140, 610)
Chloroform; aqueous	Ahuja and Spitzer (10)
	Ahuja et al. (13)
	Anderson et al. (31, 32)
	Sebborn (753)
	Tsuji (847)
	Wallace (887)
	Zagar et al. (929)[b]
Chloroform:carbon tetrachloride: isobutanol (5:4:1); aqueous	Sansur et al. (716)
Chloroform:diethyl ether:pentanol (80:20:5); aqueous	Blackmore et al. (99)

TABLE 1 (continued)

Phases[a]	Reference
Ethyl acetate; aqueous	Viktora and Baukal (873)
Ethyl acetate:heptane (3:2); aqueous	Viktora and Baukal (873)
Ethylene dichloride; aqueous	Fournier et al. (286, 287)
	Kream et al. (498)
Heptane; aqueous	Lorch (550)
Heptane:isopropanol:1 $\underline{N}$ sulfuric acid (1:1:0.08); aqueous	Kashket (453)
Hexane:ether (2:1); aqueous	Hormann et al. (413)
Isoamyl alcohol; aqueous	Manly et al. (565)
Isobutanol; aqueous	Pelletier and Brassard (651)
Isopropyl ether; aqueous	Viktora and Baukal (873)
Methyl ethyl ketone; aqueous	Avanzini et al. (48)
Sulfuric acid:ethanol (4:1); cyclohexane	James and Townsend (432)

[a]The first-named phase is recovered; the second flows to waste.

[b]Both phases recovered simultaneously.

Kashket (453) gives a drawing of a modified B4 separator used to recover an organic phase (isopropanol:heptane:1 $\underline{N}$ sulfuric acid, 1:1:0.08), discard the aqueous phase, and debubble the stream. Lorch (550) published a figure of a phase separator/debubbler arrangement made from two C4 fittings and tubing inserts which was used to recover n-heptane from an aqueous phase. Anderson et al. (31, 32) used a fairly high air bubble flow rate (2.5 ml/min) to segment a mixture of chloroform and aqueous phases as it passed through a beaded extraction coil. When a B0 separator was used, the high air rate caused excessive disturbances in the flow pattern. The problem was solved by interposing a C5 debubbler between the beaded coil and the B0 separator. A 10- to 12-in. length of Tygon tubing (0.25 in. i.d. x 0.375 in. o.d.) was attached to the upper arm of the C5 debubbler, making a smooth natural bend that extended upward vertically, open to the air. This tubing aided in the smooth release of the air bubbles without loss of liquids. The lower chloroform phase was pumped from the B0 separator, and the upper aqueous phase was led from the B0 overflow to waste through a gravity line. If this gravity feed line were too long and narrow, periodic siphoning would occur, which would disturb the liquid flow pattern.

Carter and Nickless (169) show a phase separator used to recover carbon tetrachloride from an aqueous phase. The top of the separator is open to the atmosphere; the phases enter at the top through a tube of smaller diameter than that of the separator opening. The carbon tetrachloride layer is pumped off the bottom of the separator, and the aqueous phase flows to waste through a side arm. The fitting is used with the single pump tube, double displacement flask system described in Chapter 2, Sec. IX, C. This system permits exactly matched rates of addition and withdrawal of the organic phase, resulting in virtually complete recovery of the carbon tetrachloride. The separator features small volume, 0.25 ml, and very low mixing action.

Several fittings have been designed to facilitate the reunion of the organic phase segments, thereby aiding in a clean separation and recovery. Generally these designs feature either a glass capillary, a wire insert, or a plastic insert. Browett and Moss (139, 140, 610) used a phase separator into which alternating segments of aqueous and organic (chloroform and carbon tetrachloride) phases entered through an inlet tube which projected inward and then downward, directing the stream toward the bottom of the chamber. The lower organic phase exited through a capillary tube, while the upper phase exited through a horizontal capillary tube which joined a vertical capillary tube. The upper end of the vertical capillary tube was vented to the atmosphere, and the stream from the lower end flowed to waste. Tsuji (847) shows two diagrams of micro separator fittings used to discard an aqueous phase and recover a chloroform phase. Both designs feature a capillary tip through which the alternating phases enter the separator, facilitating the formation of chloroform droplets from the mixture. Blackmore et al. (99) show pictures of three separators based on Tsuji's concepts and discuss applications of the several designs.

Valentini (857) gives a diagram of a special separator used to recover a benzene:isopropyl alcohol (9:1) phase from an alkaline saline solution. The separator consists of a T piece, slightly fattened at the junction, containing a carefully placed length of stainless steel wire. The end of the wire punctured the interfaces of the incoming stream, forming an organic phase which could be withdrawn smoothly. Manly et al. (565) used a B4 fitting containing 4 in. of coiled platinum wire (No. 28 size) to effect a recovery of an isoamyl alcohol phase from an aqueous phase. Zagar et al.

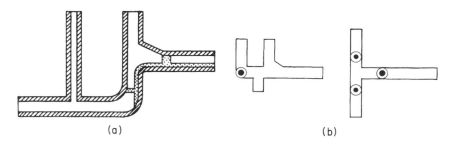

(a)   (b)

FIG. 5. Teflon strip phase separators. Solution flow is from right to left. (a) Segmented mixture of immiscible phases enters the fitting. Aqueous phase (and air bubble if present) exits upward; organic phase (in this case chloroform:carbon tetrachloride:isobutanol, 5:4:1) wets the Teflon strip and is guided downward. Immediate resegmentation of the separated organic phase is possible by adding air through the leftmost vertical tube. (b) As in (a), but the organic stream is split just past the point of phase separation. Air bubbles may then be added immediately to both exiting resampled streams. Excess organic phase flows to waste from the lower exit portal. Side view left, top view right. Reproduced from Sansur et al. (716) by permission of the authors and the Association of Official Analytical Chemists.

(929) recovered both an aqueous phase and a chloroform phase, each sent to a separate determinative step on the manifold, by passing the mixture through a separator fitting containing a twisted platinum wire that was bent so that one end of the wire pierced the chloroform globules, disturbing the surface tension and favoring their reunion.

Wallace (887) shows a special glass phase separator used to recover a chloroform phase from an aqueous phase and to send the former to a flowcell. The mixed phases enter a chamber, the top of which is open to the atmosphere. The chamber contains a carefully placed length of polyethylene tubing. The chloroform wets the tubing and is formed into medium-sized globules which are led downward into a cleanly formed layer. The separator is designed to recover all the chloroform extract. If the chloroform supply is interrupted, the separator does not deliver the aqueous phase to the flowcell, but simply stops the flow. The key to this action lies in careful positioning of the exit portals and in the use of gravity to feed the flowcell. Sansur et al. (716) describe separators which minimize holdup volume and which feature a strip of Teflon inside the fittings (Fig. 5). The Teflon leads the organic phase downward and away from the lighter aqueous phase. The latter separates upward through the top outlet. An air inlet is provided just past the point where the organic

phase has separated from the aqueous phase, allowing immediate resegmentation of the former with air. Another similar fitting permits splitting
the separated organic phase into two streams, each of which may be
immediately resegmented with air.

Sometimes it is difficult to maintain balanced flows to and from phase
separator fittings. To avoid the use of such fittings, Ganis (303) employed
a funnel containing Whatman IPS Phase Separating Filter Paper (Reeve
Angel and Co.). The mixed immiscible phases were pumped into the top of
the funnel. The organic phase passed through the paper and was pumped
to the next stage on the manifold. The aqueous layer and residues
remained in the funnel and were drawn off to waste.

A special "solvent exchanger" tube was described by Child and Caisey
(184). After oxidation, urinary steroids were extracted into methylene
dichloride on the manifold. The separated organic phase was combined
with an aqueous colorimetric reagent stream and both were then passed to
the solvent exchanger, a water-jacketed tube about 300 mm long, 15 mm
i.d., maintained at 75°C and inclined at an angle of about 35 deg. The
mixture of methylene dichloride and aqueous reagent ran down the tube by
gravity against a stream of air flowing upward through the tube. The
system was adjusted so that all the methylene dichloride boiled off by the
time it had passed about two-thirds of the way down the tube, thus
transferring the steroids into the aqueous reagent. The aqueous phase
exited at the bottom of the tube and was resampled for further treatment.
This system of transferring an analyte from an organic phase to an
aqueous phase is very similar to a technique discussed in relation to the
Continuous Digestor in Chapter 2, Sec. IV, B, 2, to which the reader is
referred.

5. Emulsions

Prevention or breaking of emulsions has been accomplished by several
procedures. Ahuja et al. (10, 13) found that the use of beaded coils caused
serious emulsions to form when a drug was extracted from an aqueous
phase into a chloroform phase. Three double mixing coils in series were
substituted for the beaded coil, and the chloroform phase was recovered
through a B0 separator. A stream of water was then added to the
chloroform stream, and the chloroform phase was again separated through
a second B0 fitting. The water wash eliminated any residual emulsions in
the chloroform extract. Valentini (856) gives a drawing which shows the
use of several T fittings in series, with or without the addition of a stream
to aid in breaking emulsions, to effect clean separation and recovery of a
benzene:isopropanol (9:1) phase from an aqueous phase. Lorch (550)
designed an AutoAnalyzer system in which free fatty acids were extracted
into n-heptane. The n-heptane phase was then washed with water in a

Teflon tube filled with siliconized beads before being passed to the phase separator. Viktora and Baukal (873) found that the dilution of the sample with aqueous sodium chloride improved phase separation and extraction efficiency when extracting a drug (diazoxide) from blood or urine samples into ethyl acetate, ethyl acetate:heptane (3:2), or isopropyl ether phases. Korf et al. (495) describe AutoAnalyzer methods for the determination of 5-hydroxytryptamine (5-HT) or 5-hydroxyindoleacetic acid (5-HIAA) in tissue extracts or fluids. In the 5-HT manifold, aqueous ammonia was added to the sample and the 5-HT was extracted into butanol:heptane (3:1); the organic phase was separated, and the 5-HT was extracted into aqueous sulfuric acid and determined fluorometrically. In the 5-HIAA manifold, aqueous sulfuric acid was added to the sample and the 5-HIAA was extracted into butanol:heptane (1:1); the organic phase was separated, and the 5-HIAA was extracted into aqueous potassium hydroxide and determined fluorometrically. All the aqueous reagents mentioned above contained 200 g/liter of sodium sulfate decahydrate. The presence of the salt facilitated the separation of the immiscible phases and, in the first of the two extraction steps, it increased the extraction of the compounds into the organic phase.

Wallace (887) noted that a haze may form in the chloroform phase after the extraction of a compound from an aqueous phase and solved the problem by building a small heater coil around the phase separator fitting to warm it.

6. Special Applications

Before leaving this topic, we mention two unusual and interesting papers on the technique of extractions. Bano and Crossland (58) described an AutoAnalyzer method of analysis for lead in samples of calcium carbonate, lime, or limestone. A liquified ion exchange resin (Amberlite LA-1, 15% v/v in heptane) was used to extract the lead from the alkaline aqueous stream. The resin phase was recovered as the upper layer from a B4 fitting, and the lead was then extracted from the resin into an aqueous buffer. The buffer phase was recovered and extracted with dithizone in chloroform, and finally the lead was determined colorimetrically. The resin solution was injected into the sample stream in small droplets through an A6 fitting and the extraction was carried out through a double mixer coil. The mixer coil used in the extraction of lead into the dithizone-chloroform reagent was wrapped in black paper to avoid photodecomposition. After passing through the manifold, the resin solution was recovered and saved for subsequent regeneration and reuse.

Nyberg (631) discusses the theory of continuous flow, dye complex (ion pair) determinations of aliphatic primary, secondary, tertiary amines and quaternary ammonium compounds, using methyl orange as the dye reagent. Equations are given to calculate from partition coefficients whether a given extraction will be complete on the manifold. Techniques are described using selective masking reactions on the manifold prior to developing the color that will differentiate between the classes of compounds mentioned above. Referring to an excellent article by Schill (732), which covers the theory of manual dye complex determinations, Nyberg shows how to translate the theory into practice on the AutoAnalyzer.

H. Separation of Solids

In addition to glass trap fittings (as mentioned above in Sec. II, A, and in Chapter 2, Sec. XII, A), in-line filters may be used to remove unwanted solid materials from flowing streams. A common filtration material is glass wool, lightly packed in a short section of glass transmission tubing or placed directly in one or more of the glass manifold fittings (84, 498, 785). Gregory and Van Lenten (348) show a glass in-line filter containing a sintered Teflon disk to support the glass wool. Blackwell (100) made in-line filters from the barrels of 5-ml disposable syringes lightly packed with fine glass wool. Each filter was conditioned in the manifold waste stream about 30 min before it was used to replace an old filter.

Several other filtration devices have been used in continuous flow systems. Michaels et al. (596) employed an AutoAnalyzer in a study of the dissolution rates of tablets. A sintered glass tip was inserted into the sampling line running to the manifold to filter the withdrawn dissolution medium. The filter was strapped onto the basket assembly of a disintegration apparatus. Schroeter and Hamlin (740) used a pump, a three-way valve, a fritted glass filter, and a spectrophotometer to study the dissolution rates of tablets and capsules. In its normal position, the valve returned the sampled stream directly to the dissolution vessel. At 1- or 2-min intervals, a timer actuated the valve for 15 to 20 sec, sending the stream through the filter to the flowcell and then back to the dissolution vessel. The valve then returned to its normal position. This discrete filtering action was used in preference to continuous filtration and monitoring to avoid clogging the fritted glass filter. Burns and Hansen (155) show a diagram of a holder which permits using Millipore filters in AutoAnalyzer manifolds. Their arrangement includes a system to backflush the filter, cleansing it between samples. Reid and Wise (684) employed Whatman filter tubes on all reagent lines to prevent particles from the reagents entering the manifold.

Fittings for the separation and optional recovery of solids have been reported in several papers dealing with blood analysis. Grunmeier et al. (351) and Cohen (188, 189) show figures of fittings designed to decant agglutinated blood cells from flowing streams. Albisser and Leibel (15) diluted blood samples with an isotonic diluent (0.15 $\underline{M}$ lithium nitrate), producing a density gradient in the flowing stream which caused rapid settling of the blood cells by gravity.   The cells were then separated by passing the stream through a decantation fitting tilted to promote with-drawal of the relatively cell-free fraction while discarding the cell-rich fraction.   Lithium nitrate was used because it provided the reference standard for a flame photometer and did not interfere with the desired electrolyte determinations. The density gradient method of cell separation was believed to be more efficient than dialysis and gentler than either filtration or centrifugation, rupturing fewer cells.

I. In-Line Columns

We now consider columns placed in-line on the manifold and used either to remove unwanted constituents from the flowing stream or to effect a chemical reaction.   Columns used to achieve chromatographic separations are treated in Chapter 5.

1. Selective Removal of Unwanted Species

Ion exchange resins have been put to several interesting tasks. Conca and Pazdera (193) used such a column to remove interferences in the determination of streptidine. The resin was fractionated, and the 25- to 40-μm particle size range was placed in a jacketed column maintained at 30°C by a circulating water bath. Each sample was added manually to the column, and the column was then connected to the Pump.

Gaddy and Dorsett (297) reported an AutoAnalyzer method for the analysis of free acid in the presence of hydrolyzable ions such as aluminum. An ion exchange column was placed in the manifold to remove these interfering ions. It was a 4-ft length of Tygon tubing, 3 mm i.d., filled with AG 50 W-X8 or 50 W-X4 resin. The shape of the column was not critical. Essentially the same results were obtained when using either straight lengths of glass (3 mm i.d.) or Tygon tubing in vertical or horizontal positions or when using tubing formed into coils 5 in. in diameter. Each sample was aspirated for 15 sec, followed by a wash of 8 min 45 sec.

In-line ion exchange columns have also been used to remove inter-ferences in several automated determinations of sulfate (606). Lazrus et al. (525) placed the resin in a column made from a 6-in. length of glass tubing. Gales et al. (299) removed cations by passing the stream through a 12-in. length of 0.1875-in.-i.d. tubing containing Dowex 50 W-X8 resin. Wurzburger (923) used an AutoAnalyzer to analyze for sulfur in plant tissue digests and employed a 15-cm length of 0.5-cm-i.d. tubing, which was mounted vertically on the manifold system, filled with cation exchange resin, and plugged at both ends with glass wool. The column packing was renewed daily.

Dowd et al. (232) analyzed for calcium pantothenate in multivitamin preparations. The sample was pumped through a column containing a layer of Dowex 50-X-4 resin and a layer of Florisil. The column removed dyes and other interfering materials and was repacked every 2 days; its dimensions were 15 cm x 0.64 cm. Samples were analyzed at the rate of 10 per hour; 45 sec of sampling and 5 min 15 sec of wash were used. About 45 min elapsed between the aspiration of a sample and the appearance of the corresponding recorded peak. This column constituted a very effective "cleanup" since many different formulations could be analyzed. A similar column was also used by Albright (16) in the automated analysis of calcium pantothenate in multivitamin preparations.

Urbanyi and O'Connell (852) published an AutoAnalyzer method of analysis for hydralazine hydrochloride, reserpine, and hydrochlorothiazide in combination in drug formulations. An ion exchange column was placed in-line on the manifold to retain hydralazine and reserpine while passing hydrochlorothiazide. The resin was sized to the 40 to 60 mesh range, and was placed in a 150-mm length of 1-mm-i.d. Solvaflex tubing by plugging one end of the tube with a small glass wool plug and then drawing a water slurry of the resin through the tubing with vacuum. The column lasted for about 150 to 200 assays before replacement was necessary. The use of ion exchange columns to retain interferences while passing the desired analyte appears to be a very powerful separation technique whose potential has only begun to be explored in continuous flow analysis.

Fleet et al. (282) described a device to remove water and carbon dioxide from a flowing gas stream. After suitable reactions on the manifold, oxygen was liberated in the sample stream. A stream of nitrogen, pumped at 6 ml/min, swept the oxygen from the liquid stream through two gas-liquid separators and then through a U tube, 14 cm long, 4 mm i.d. The tube was three-quarters filled with soda-lime asbestos (Carbosorb, 12-30 mesh with self-indicator) and the remainder was filled with dry calcium chloride.

2. Chemical Reactions

We now turn to the use of on-line columns to effect chemical transformations on the manifold. Blaedel and Todd (111) passed a stream containing α-amino acids through a reaction tube packed with 50 to 60 mesh copper phosphate. An equivalent amount of Cu(II) ion was liberated and was then measured by continuous flow polarography.

Several authors have described the use of in-line columns to reduce nitrate ion to nitrite ion. The nitrite ion thus produced may be used to diazotize an aromatic amine, which then may be coupled with a suitable reagent (e.g., N-1-naphthylethylenediamine) to yield an intensely colored dye, thereby permitting a colorimetric determination of nitrate ion. O'Brien and Fiore (633) constructed a reduction column from a 12-in. length of transmission tubing filled with granular zinc and plugged at both ends with glass wool. The column was connected to the rest of the manifold via sleeve fittings and Teflon nipples. Gases generated in the reaction were vented to waste along with air bubbles through a T fitting placed directly after the reduction column. The stream was then resegmented and passed to the colorimetric portion of the manifold. A Sampler I was operated at 60 samples per hour, with two wash cups placed between each sample. If nitrite ion itself were present in the samples, its analysis could be obtained by omitting the reduction column, permitting suitable corrections to be made in the sample nitrate values. Litchfield (544) also used a zinc column to reduce nitrate to nitrite. The column was placed between two 2-way stopcocks. When both stopcocks were thrown to their alternate positions, the flow stream bypassed the reduction column through an unpacked tube, allowing the determination of nitrite only. Litchfield used a 30-cm length of transmission tubing; a chip of zinc was pushed to the center of the tube and it was then filled by adding zinc chips from both ends. The column was washed free of air bubbles but was left filled with reagents at the end of the day to prolong its life. The column also reduced organic nitrates, e.g., nitroglycerin. If present, such compounds would cause positive interferences, and they were removed from the samples by a preliminary manual extraction.

Cadmium has also been used for in-line reduction of nitrate. Brewer et al. (133) placed cadmium filings retained by a 20-mesh sieve in a single mixing coil. This coiled column gave between 4.6 and 77.0% reduction of nitrate in samples of seawater and up to 92.1% reduction of nitrate in distilled water, depending on the concentration of EDTA in the stream. The EDTA was used to maintain the reactivity of the cadmium filings. Grasshoff (342) used cadmium of 40 to 60 mesh in a U-shaped glass tube, 14 in. long, 3 mm i.d. Henriksen and Selmer-Olsen (390) reduced nitrate to nitrite by passing the stream through a tube (140 mm x 4 mm) containing "copperized cadmium" held in place with glass wool plugs. This column lasted 2 to 3 months in regular use. Bradshaw and Spanis (127, 128) prepared columns by pretreating cadmium filings with copper sulfate solution and placing the prepared filings in a disposable Pasteur pipet. One column lasted about 2 weeks.

J. Irradiation

Radiant energy in the ultraviolet and visible regions has been used to effect several useful photochemical transformations in continuous flow manifolds. Skeggs and Hochstrasser (775) increased the sensitivity of a glucose determination by passing a stream through a double mixing coil placed between two 4-W fluorescent lamps (General Electric F4T5/CW). The lamps directly contacted the coil, and the whole was wrapped in aluminum foil. Larsson and Samuelson (522) also used a double mixing coil between two 4-W fluorescent lamps, and Lee et al. (526) used a Technicon fluorescent sensitizer lamp in manifolds designed to monitor column eluates for sugars.

Organic nitrogenous compounds were oxidized to nitrite and nitrate ions in a manifold described by Afghan et al. (6). A 550-W ultraviolet photochemical lamp was placed inside two quartz coils, 5 in. in diameter, made from 3-mm-i.d. tubing of 0.6 mm wall thickness. The sample stream was acidified and passed through the first coil; it was then made alkaline and passed through the second coil.

Grasshoff (344) shows a quartz coil apparatus built around a 900-W lamp, which was used to digest organic phosphates.

A mercury pen lamp placed in the center of a silica coil enabled Heinicke et al. (382) to study the effect of ultraviolet light on enzymes.

Several authors have reported the use of irradiation in continuous flow methods for the analysis of cyano compounds. Dowd et al. (231) developed an AutoAnalyzer method for cyanocobolamine. They show a time-delay coil adapted for photolysis with a light bulb inside and a reflective surround. In the presence of sulfuric acid, light catalyzed the replacement of the cyanocobolamine cobalt-complexed cyano group by a sulfate group. The liberated cyanide was then determined colorimetrically. Although this irradiation apparatus worked well for clear solutions, when turbid or nontranslucent solutions were tested, only the layers near the lamp were affected. The light and time-delay coil assembly was then replaced by a Continuous Digestor module. The glass helix was illuminated by a series of floodlights, and the thin film of liquid permitted efficient photolytic catalysis. Later, Love and McCoy (553) reported improvements in the delay coil/light apparatus that eliminated the need for the Continuous Digestor in the cyanocobolamine determination. Quartz-halogen lamps, which gave much more intense light than floodlamps, were used with two 40-ft delay coils. A cooling fan allowed the lamps to be placed close to the coils. Goulden et al. (339) analyzed for nanogram quantities of simple and complex cyanides using the ultraviolet irradiation apparatus of Afghan et al. (6) described above.

Hussey et al. (422) used ultraviolet irradiation to develop the color for analytical measurement in an automated analysis of diethylstilbestrol in tablets. The solution was pumped through capillary tubing of 6.0 ml capacity, which was bent into the shape of a grid, 160 mm long per turn. The grid was backed with an aluminum reflector to intensify the illumination. A figure of the grid appeared in their paper.

K. Distillation

In this section, we shall consider continuous flow separation of volatile materials as gases from liquid streams. Although heat is usually required in the distillation step, several methods operate successfully at room temperature. For example, we have already mentioned, in Sec. II, I, 1, the separation of gaseous oxygen from a liquid flow stream (282). The use of the Technicon Digestor module in continuous distillation methods is discussed in Chapter 2, Sec. IV, B, 3, to which the reader is referred.

1. Carbon Dioxide

In 1960, Skeggs (768, 770, 774) published a method for the analysis of serum samples for carbon dioxide content. Gaseous carbon dioxide was released from the sample stream and separated through a gas-liquid separator fitting. The gas stream was then segmented into a colorimetric reagent, such as buffered phenolphthalein solution, and the carbon dioxide content was measured colorimetrically. Hales et al. (365) modified Skeggs' gas-liquid separator to incorporate a reservoir. The modification avoided the "greatly varying volume of gas encountered when the original straight-through waste line was used." In addition, the inner walls of the separator were coated with silicone grease to reduce foaming.

Skeggs' procedure for carbon dioxide opened an interesting pathway for the determination of many classes of compounds. If a reaction is available in which the analyte plays a controlling role in the formation of carbon dioxide, one may design the necessary manifold to obtain the carbon dioxide and then couple it to Skeggs' determinative step. Thus, Gerke et al. (324) analyzed for an antibiotic by measuring the amount of carbon dioxide released by bacteria treated with the antibiotic samples. Jones and Palmer (440) used a vacuum-jacketed separator described by Shaw and Duncombe (758) to recover the carbon dioxide liberated in the determination of ampicillin and carbenicillin. The interior walls of the separator were treated with silicone grease, special care being taken at the point where the gas-liquid stream entered the trap to ensure that the bubbles were not lost by flowing directly to waste. Schaiberger and Ferrari (729) analyzed for L-lysine via enzymatic decarboxylation and subsequent determination of the released carbon dioxide. Knowles and Hodgkinson (493) described a determination of serum oxalic acid using enzymatic decarboxylation, followed by a modified Technicon method for measuring the liberated carbon dioxide. They designed a special gas-liquid separator made of Perspex which minimized the formation of aerosols at the point of separation, reducing the carryover of liquid phase into the gas phase. After the initial separation, the gas phase bearing the carbon dioxide was pumped through a filter containing a Whatman glass paper disk (GF/C grade) to remove residual traces of aerosol.

2. Nitrogen and Nitrogenous Materials

Prall (676) determined the nitrogen content of metallic uranium, distilling the analyte as ammonia on the manifold, and shows a drawing of the distillation apparatus, which was constructed from a 100-ml Pyrex drying tube. The sample was made alkaline and then passed into a U-shaped inlet tube, which was wound with felted asbestos-insulated No. 22 nichrome V wire, 14 Ω resistance, connected to a variable voltage device set at 32 V. The heated inlet tube brought the sample stream to the distillation temperature and the ammonia vapors were trapped in boric acid solution which was pumped through a reservoir at the top of the distillation fitting. Keay and Menage (459) determined nitrogen (as ammonia) in feces and soil samples. They show a figure of a steam distillation apparatus designed for use with an AutoAnalyzer. The sample is made alkaline and pumped into a 2-ft steel coil placed in a specially made heating bath set at 116°C. At the coil exit, a stream of air is added, forcing the gas-liquid mixture into a series of splash heads. The liquid fraction is pumped to waste; the air-steam-ammonia phase passes through the splash heads and meets a stream of hydrochloric acid in which the ammonia is dissolved. The mixture passes down a condensing column to a phase separator, and the acidic liquid phase is recovered for a colorimetric finish. The Sampler was operated at 30 samples per hour with a sample to wash ratio of 1:1. Hanawalt and Steckel (373) describe how to construct a continuous vacuum still from standard Technicon coils and traps. Ammonia is distilled away from interferences (sediment, color, and heavy metals in soil solutions) and then determined colorimetrically.

Ayers (50) steam-distilled alkaloids from tobacco and smoke samples and then measured the alkaloids via ultraviolet spectrophotometry. A Technicon steam distillation unit was used, a figure of which appears in Ayers' paper. The aqueous alkaline sample stream passed through a glass tube heated in an oven where a portion of the sample stream was converted to steam; the steam and remaining sample stream then passed into the steam distillation unit. Arend et al. (42) reported a method for the steam distillation and subsequent determination of diazotizable aromatic amines.

On the manifold, kynurenine and acetylkynurenine were treated with alkali and heat, producing o-aminoacetophenone. The latter was steam-distilled by sending the stream into a Teflon coil in a Heating Bath set at 145°C. The solution and steam were passed into a Teflon still head, and the vapor phase was routed to a water-cooled condenser. The condensed liquid phase containing the amine was then recovered and monitored colorimetrically. The system was operated at 12 samples per hour.

3. "Fluoride"

Manly et al. (565) distilled "fluoride," probably as silicon tetrafluoride. The analyte was trapped from the vapor phase, extracted, and determined colorimetrically. Mandl et al. (564) give a photograph and drawing of a microdistillation device used in an AutoAnalyzer method for fluoride. Solutions of ashed or alkali-fused samples are mixed with sulfuric acid, segmented with air, and passed into a Teflon coil heated at 170°C. The "fluoride" and water vapor distillate are separated from the acid and ashed solids via the microdistillation column and a liquid trap. The recovered analyte is then determined colorimetrically. The samples were analyzed at the rate of 20 per hour. Marten and Grady (579) show a picture of a distillation column, designed for use with AutoAnalyzer manifolds, featuring an evacuated outer jacket. Holy (410) described a manifold in which the acidified sample was brought to 120°C in a Heating Bath, and the hot exit mixture was fed into a glass vessel. The vapors rose into a scrubber and were trapped and mixed with the color reagent while the spent sample liquid phase was drawn off the bottom of the vessel and sent to waste. Catanzaro et al. (171) show a picture of a glass continuous distillation apparatus featuring an absorption chamber in which the colorimetric reaction takes place. Wahl and Auger (881) give a drawing of a thermostatically controlled aluminum block heater used to flash-distill "fluorine." It contained a 15-ft Teflon coil and was maintained at 190°C. Copper dust was placed around the coil to act as a heat transfer medium. Oliver et al. (635) added sulfuric acid to fluoride samples, distilled the samples, and monitored the sulfuric acid and hydrofluoric acid recovered in the distillate using continuous flow pH and fluoride electrodes (see Chapter 3, Sec. II, D).

4. Phenols

Continuous distillation of phenols has been described by Friestad et al. (292, 293). The stream to be distilled entered a single mixing coil wrapped with heating tape and adjusted to about 130°C; it then passed to a Technicon microdistillation column also wrapped with heating tape. The heating tape power was controlled by a variable transformer. The single mixing coil was used in preference to the long, large-diameter Teflon coil supplied by Technicon, because broader, lower recorder peaks were obtained when the latter coil was used. Ott et al. (639) analyzed for carbaryl (as 1-naphthol), utilizing steam distillation equipment similar to that of Friestad et al. (292, 293). The distillation apparatuses reported by Catanzaro et al. (171) and Marten and Grady (579), discussed above in connection with fluoride analysis, were also suitable for the distillation of phenol.

5. Cyanide and Cyano Compounds

Love and McCoy (553) photochemically released cyanide from cyanocobolamine and swept hydrogen cyanide from the sample stream with a flow of nitrogen, discarding the spent sample liquid. The hydrogen cyanide was then absorbed in the next liquid reagent. This arrangement simultaneously eliminated nonvolatile interferences and concentrated the cyanide. A drawing of the glass fitting used in the distillation appears in their paper.

Goulden et al. (339) describe manual and automated distillation methods for simple and complex cyanides. In the manual procedure, hydrogen cyanide is distilled into a trap using a packed column. Sodium hydroxide solution, which absorbs the cyanide, is constantly recirculated through the column by a Technicon Proportioning Pump. In the automated method, a 20-channel Carlo Erba proportioning pump was used, along with a Technicon Sampler II, Colorimeter, and fittings. After photochemical liberation of free cyanide from complexes in an irradiation coil, the stream passes through a continuous distillation apparatus wrapped in 22-gauge heating wire and a continuous gas absorption column packed with 0.125-in. helices. Hydrogen cyanide is distilled over and then is absorbed in sodium acetate solution continuously flowing down the gas absorption column. The analysis is completed via an automatic colorimetric determination. To determine simple cyanides, i.e., those readily volatilized by acidification, the irradiation coil is bypassed on the manifold. Goulden et al. give drawings of the distillation and absorption apparatuses and discuss their parameters in terms of the system response versus the fraction of sample vaporized, the air flow through the distillation tube, and the height of the absorption column.

The apparatus described by Holy (410), and discussed above as used in fluoride analysis, was also employed in cyanide distillation.

6. Sulfur and Sulfureous Materials

The analysis of sulfur (as sulfate) in leaf sample digests, feces, and soil extracts is described by Keay et al. (460). The sulfate is reduced to hydrogen sulfide with hydriodic acid, formic acid, and hypophosphorus acid, and then the stream is passed through a special coil maintained at $120^{\circ}C$. At a point in the coil where sufficient reduction has occurred, a stream of nitrogen is added, sweeping the gas-liquid mixture into a special condenser-separator fitting, which is shown in their paper. After separation from the spent liquid phase, the gas phase is passed over a slowly flowing stream of sodium hydroxide solution, and the stream of sodium sulfide thus formed is withdrawn for a turbidimetric finish. The condenser-separator fitting features a water jacket to cool the upward flowing gas stream, thus condensing out the steam, and a separate, water-jacketed coil to cool the condensate before it is pumped to waste. Keay et al. describe the lengths and internal diameters of the coils used in the $120^{\circ}C$ bath that gave the best recoveries. The analysis rate was 30 samples per hour.

Prescott (678, 679) devised an AutoAnalyzer method in which the analyte, lincomycin, was hydrolyzed to volatile methanethiol. The methane-thiol was separated from the hydrolyzed sample stream through a B1 gas trap and then mixed with a colorimetric reagent. Reagents and sample solutions were degassed when necessary to avoid upsetting the stream flow at the gas trap.

7. Alcohol

Sawyer and Dixon (726) led samples of beer through a bath held at $95^{\circ}C$; a stream of nitrogen was added to the liquid stream inside the bath, sweeping the alcohol fumes into a separator fitting. The sample residues drained to waste and the nitrogen-alcohol stream passed to a condenser. A stream of water was added to the condenser to help trap the alcohol, and a colorimetric determination was carried out on the condensate. In a similar procedure for alcohol in beer, Sawyer and Dixon (725) used a flow of air rather than nitrogen to sweep the alcohol from the heated coil to the separator fitting.

Davies et al. (219) show a glass distillation head used to distill alcohol away from interferences in an AutoAnalyzer method for the analysis of aqueous flavor bases. The reagent which absorbed the distilled alcohol was also the colorimetric reagent, and it was pumped from the distillation apparatus to the Colorimeter.

8. Volatile Aldehydes and Ketones

Duncombe and Shaw (234) presented an AutoAnalyzer method for volatile aldehydes and ketones. The samples were automatically steam-distilled and colorimetrically determined. Figures of several interesting fittings appear in their paper, including a miniature glass "cyclone" used to separate liquid and gas phases under conditions of high gas flow rates and a jacketed glass helix, used vertically, in which the sample liquid flows downward while a gas stream passes over the liquid surface to remove volatile analytes.

L. Gradients

1. Method and Mathematics

Ryland et al. (712) described a technique to produce a solution of gradually increasing concentration, using a Proportioning Pump and a vessel containing a mechanical stirrer. The Pump is provided with two pump tubes of equal pumping rates. The vessel is filled with a solvent or solution devoid of the material whose concentration is to be increased. One of the pump tubes is arranged to deliver a solution of the material into the mixing vessel; the other is arranged to remove mixed liquid at the same rate so that the volume of solution in the vessel remains constant. Assuming that mixing is instantaneous, the concentration of the material in the vessel may be found from Eq. (1):

$$c_t = c_i \left[1 - \exp(-rt/v) \right] \tag{1}$$

where c_t = concentration at time t

c_i = concentration of solution being pumped into vessel

r = pump rate

v = volume of liquid in vessel, a constant

At the start of the experiment, t equals zero, rt/v equals zero, and the exponential term equals 1, giving c_t equals zero, as expected. As t increases, the exponential term decreases, eventually reaching zero as t reaches infinity. Thus, c_t starts at zero and approaches c_i, the terminal concentration in the vessel.

To understand the parameters involved in such an experiment, it is of interest to solve Eq. (1) for time. Let x be the fraction of the terminal concentration c_i in the vessel at time t_x. Then,

$$c_{t_x} = xc_i$$
$$x = 1 - \exp(-rt_x/v)$$
$$\ln(1 - x) = -rt_x/v$$
$$t_x = -(v/r) \ln(1 - x) \tag{2}$$

Equation (2) shows that t_x is independent of concentration, depending only on the pump rate and the volume of liquid in the vessel.

For example, let r equal 2.39 ml/min and v equal 97 ml. To calculate the time required for the concentration to increase to one-half its terminal concentration, x equals 0.5, and

$$t_{0.5} = - \frac{97 \text{ ml}}{2.39 \text{ ml/min}} \ln(1 - 0.5) = 28.1 \text{ min}$$

For the time required for the concentration to reach 95% of c_i, x equals 0.95 and $t_{0.95}$ equals 121.6 min. Thus, for this combination of volume and pump rate, almost the entire gradient range will be covered within 2 hr.

Equation (1) applies when starting from a concentration of zero and working up to the terminal concentration. In some cases, one may wish to start with the highest concentration in the gradient vessel at time zero and dilute downward, letting the concentration approach zero as time approaches infinity. In such experiments the equation given by Schwartz and Bodansky (747) may be of use:

$$c_t = c_0 \left(\frac{v - r}{v} \right)^t \tag{3}$$

where c_t = concentration at time t

   c_0 = initial concentration in vessel when t = 0

   v = volume in vessel, a constant

   r = pump rate

Again, the pump rate into the vessel equals the pump rate out. Equation (3) is apparently an empirical relationship, for neither the units in the parenthetical term nor those in the power term reduce to a dimensionless value. Nevertheless, the equation was verified experimentally by Schwartz and Bodansky through continuous measurement of the absorbance of a gradient stream formed as described. These authors stress the importance of calibrating the two pump tubes and ensuring that a matched pair is used.

2. Applications

The gradient concentration technique has been put to several interesting tasks. In one of the earliest reports, Pagano et al. (643) used continuous dilution to obtain complete dose-response curves in automated microbiological assays. Roodyn (690, 691) used a gradient mixing system to permit measurement of widely varying concentrations of enzymes in automated multiple-enzyme analyses. Roodyn and Maroudas (693) carried out similar experiments and described the calibration of a gradient dilution system by substituting a colored solution for the enzyme solution and following the absorbance of the gradient stream with time. Posen et al. (671) used continuous gradient dilution in automated studies of enzyme reaction kinetics. Gradients have been used to evaluate the interference of urea on

alkaline phosphatase reaction kinetics (96), of other ions on the determination of magnesium (461), and of other amino acids on the determination of
methionine (854). Vargues (866, 867) employed continuous dilution in
complement-fixation blood tests and described the required mathematics.
Gradient elution of chromatographic columns may be accomplished using
continuous dilution techniques; this application is discussed in Chapter 5.

Gehrke and co-workers have written several articles on an elegant use
of gradients: the optimization of reagents used in continuous flow analyses
(314-318, 854). In this procedure, one selects a reagent whose concentration is to be optimized and sets up a gradient dilution system to supply an
increasing concentration of this reagent to the manifold. All other reagents
are pumped in the normal fashion and a steady supply of a standard or
sample solution is fed into the manifold sample line. The recorded trace
documents the effect of the slowly increasing reagent concentration on the
efficiency of the manifold chemistry. Generally one observes the time
when the strongest response appears on the recording and then calculates
the corresponding concentration of the reagent using Eq. (1). The delay
time, that is, the time from the aspiration of the reagent to the appearance
of the recorded output, must be accounted for in the calculations. The
technique of Gehrke et al. deserves careful study, for it enables one to
obtain parameters in a matter of hours that otherwise would require days
of solution preparation and data reduction. An additional example of
reagent optimization by gradient dilution is given by Law et al. (524).

Separate gradient-producing devices have also been used with Auto-
Analyzers (63). Himoe et al. (398) show a "gradient-forming device"
supplied by Glenco Scientific, Inc., Houston, Texas. The device features
two liquid chambers connected at their deepest part by a tube and separated
by a stopcock. One of the vessels is provided with a magnetic stirrer to
form the gradient. A solution flows from the first chamber through the
tube and stopcock and is mixed with the other solution in the second
chamber. The mixture exits from the second chamber to the Proportioning
Pump. The device is jacketed to afford temperature control. It created a
linear gradient and was believed to be less prone to error than the use of
the Proportioning Pump to deliver and withdraw solutions from a gradient
vessel. Tappel and Beck (821) give a photograph of a two-chamber gradient
device designed to be mounted on a common laboratory magnetic stirrer
drive. It features a surrounding bath for cooling the gradient liquids.

Lindquist (539) discusses the use of gradients to measure the effect of
interfering substances in automatic colorimetric analyses, using a device
which relies on siphoning to form the gradient. Vessel A is filled with a
concentrated solution of the interfering substance and is connected via a

capillary siphon to vessel B. Vessel B is provided with a magnetic stirrer to mix the gradient solution, and its cross-sectional area is identical to that of vessel A. As the Proportioning Pump withdraws solution from vessel B, the solution in vessel A siphons over into vessel B, producing the gradient. If no volume change occurs due to mixing, the gradient may be calculated by

$$f_b = kt/2v$$

where f_b = proportion by volume of interfering substance in vessel B

   k = measured pump rate out of vessel B

   t = time from beginning of siphoning

   v – initial volume in vessels A and B

Burns (153) reported a device for creating a salt solution gradient which was used in osmotic fragility studies. The system included circuitry to obtain the derivative of the recorded curve, a programmer to create a fresh gradient for each sample, and the appropriate connections to the AutoAnalyzer Recorder and Colorimeter. Schematics and stream flows are given in Burns' paper.

Eichler (237) developed the use of gradients in automated acid-base titrations and discusses the associated mathematics, especially with regard to linear gradients. A Colorimeter was employed to detect the start of the titration and the end point.

Brooks and Olken (137) used a manually produced gradient to determine the pH yielding the optimum fluorescence signal at fixed excitation and emission wavelengths. A buffered standard solution was aspirated until the steady-state reading was obtained from the fluorometer. A 10 $\underline{N}$ solution of sodium hydroxide was then added by drops to the buffered standard solution. Timed portions of the flowcell effluent were collected and labeled for subsequent measurement of pH and correlation to recorder reading.

Runck and Valeri (701) described a device which produced a gas phase gradient solution. Two gas syringes, driven by a reversible motor, formed a linear gradient concentration of oxygen in nitrogen. The mixed gases were fed to a glass tube, 14 in. in length, 5 mm i.d., where blood samples were continuously equilibrated with the gas stream. The liquid phase then passed through the remainder of the manifold for the determination of oxygen in the sample. The system was used to measure the rates of oxygenation and deoxygenation of red blood cells.

III. TUBING CHARACTERISTICS

A. Conditioning Tubing

Several authors recommend conditioning or breaking in new transmission tubing before use. White et al. (674) noted that new tubing may
break up bubble patterns. They conditioned new tubing by cleansing it with
water containing 2 to 3 drops of Brij-35 per liter and then "milking" it by
seesawing the tubing over a table edge. Ambrose (25, 26) soaked new
Tygon transmission tubing 24 hr in 0.07% Brij-35 solution to remove loose
plasticizers and other materials causing interference in a fluorometric
determination. After the new tubing was installed, 0.07% Brij-35 solution
was pumped through the manifold for 3 hr before the analysis of samples
was begun. Gaunce and D'Iorio (312) wanted to use Triton X-100 to improve
the flow pattern in a manifold but found that it caused precipitation of one
of the reagents (alkaline copper(II) sulfate solution). However, a good flow
pattern could be obtained by rinsing all lines with a Triton X-100 solution,
and then distilled water, before each run. Zak and Epstein (934) siliconized
manifold tubing with a water-soluble silicone.

B. Effect on Sample Interaction and "Noise"

The wash characteristics of a manifold may be markedly influenced by
the type of transmission tubing used in its construction. As one might
expect, the solvents and reagents employed in the manifold flow stream
behave differently in different tubing materials, and there is no concensus
of opinion in the literature as to the "best" tubing material. Thus, glass
tubing has been found to be superior in wash characteristics to plastic (12,
880); mixing coils made of Tygon have given better wash and less "noise"
than glass coils (277); Teflon lines and coils have been recommended
rather than glass (551); and Kel-F tubing has been found to be better than
glass for organic, alkaline solutions (702). We may draw two conclusions
from these observations. First, it is a mistake to regard any type of
tubing as totally inert and without effect on manifold wash characteristics.
Therefore, one should try several materials, choosing whichever seems
to give the best performance. Second, after selecting satisfactory transmission tubing materials, one should record their composition in the
methodology so that other laboratories will not make an incorrect choice
when using the procedure.

Reid and Wise (684) conducted experiments showing that the use of
small-bore tubing between modules improves the system's wash characteristics. The flow stream velocity was about ten times faster in 0.5-mm-i.d.
polythene tubing than in 1.65-mm-i.d. Tygon tubing; the pressure drop
increased over 100-fold. The reduced time lag between modules reduced
diffusion in the lines, leading to improved wash.

C. Adsorption-Desorption and Permeability

Another problem associated with manifold transmission tubing is the adsorption and desorption of materials in the flow stream. Auerbach and Bartchy (45) note that quaternary ammonium compounds such as cetyl-pyridinium chloride are strongly adsorbed on glass and other surfaces. Linearity curves obtained from the readings of standard solutions sent through the AutoAnalyzer system were usable but did not pass through the origin because of this adsorption.   Linton et al. (542) analyzed for benzalkonium chloride using a method based on that of Auerbach and Bartchy (45) and noted the same effect. Bittner and Manning (97) used glass transmission tubing between a Heating Bath and the flowcell in a glucose analysis because the colored complex formed in the determinative step was adsorbed on standard transmission tubing. Klein et al. (484) mention adsorption-desorption problems with plastic tubing and recommend the use of glass tubing in their automated acid phosphatase determination. Garry and Owen (309) and Hoffman et al. (406) describe AutoAnalyzer methods for ascorbic acid using the colorimetric reaction with indophenol. Glass transmission tubing was used after the color complex was formed to avoid the adsorption of the complex on plastic tubing. In designing an AutoAnalyzer method for the determination of acid normality in uranyl nitrate solutions, Prall (675) initially attempted to use bromocresol green as a reagent in a colorimetric measurement of pH. The dye became adsorbed on the manifold tubing, and when the pH of the stream went past those values corresponding to the acidic or basic forms of bromocresol green, the dye desorbed into the stream, giving false peaks. Prall then turned to the use of a continuous flow pH sensor, eliminating the need for the dye.

Phenol seems particularly prone toward adsorption on plastic tubing. Fishman et al. (279) recommend the use of "aged" transmission tubing or glass tubing to minimize phenol adsorption. Horn (414) found that consider-able amounts of phenol may be adsorbed from aqueous $0.1 \underline{N}$ hydrochloric acid solutions into Tygon tubing. The tubing did not appear to become saturated. Adsorbed phenol was released into the stream during the wash portions of the cycle. Phenol was adsorbed at pH 2 and 7, but no adsorption was detected at pH 12. Therefore, Tygon could be used only at the higher pH, glass being required at lower pH.

The adsorption of other materials has been minimized by judicious selection of pH. Sodergren (783) developed an AutoAnalyzer procedure for the analysis of detergents with methylene blue reagent. A solution of $0.01 \underline{M}$ potassium dihydrogen phosphate was supplied to the Sampler module wash box to avoid adsorption of surface-active materials onto plastic and glass tubes. Prall (677) suspected that thorium was being adsorbed onto Solvaflex tubing from aqueous solutions. To test for adsorption, Prall aspirated a solution of thorium into the system until it reached steady state and then aspirated a cup of water wash followed by a cup of 1 $\underline{M}$ nitric acid. A small peak appeared, rising rapidly, then tailing off, as the thorium was desorbed. In addition, the steady-state value from an acidic solution was higher than that from a neutral solution. Prall solved the problem by making all sample solutions 1 $\underline{M}$ in nitric acid. Additional references to adsorption of materials on manifold tubing can be found in Table 2.

Adsorption or desorption in pump tubes is especially troublesome because the choice of pump tube materials is so limited. Voss (875) found that nonpolar pesticides contaminated Tygon or Solvaflex pump tubes, causing severe tailing and carryover. Briscoe et al. (134) state that most plastics adsorb metal dithizonates strongly. Acidflex pump tubes would equilibrate after about 10 min, whereas Solvaflex removed all the zinc dithizonate from a reagent solution for several hours while pumping. Goodall and Davies (335) found iodine to be "highly soluble" in Tygon tubing and used glass to glass butt joints wherever possible along with polythene transmission tubing. The iodine-bearing stream was pumped with a silicone rubber pump tube. Loss of iodine into the pump tube was greatest during the first 30 min of operation. After that period, a slight loss continued to occur by diffusion through the tube wall.

Knowles and Hodgkinson (493) reported that polythene and polyvinyl chloride tubing were permeable to carbon dioxide in solution and used glass tubing wherever possible. However, carbon dioxide was not lost through polythene tubing from solutions of pH 3. Polythene wrapped with adhesive aluminum tape could also be used for short connections on the manifold.

D. Operational Hints

Many solvents or reagents attack plastic transmission tubing. Glass tubing has been recommended to handle ether (590), methylene chloride/ isopropanol (161), or reagents containing perchloric acid (314). Glass or Teflon tubing has been used for 85% acetonitrile (185) or amyl acetate (304), and Teflon inserts in plastic tubing have been used for isopropyl ether (804).

Several useful hints on making tubing to tubing connections have appeared. When Tygon tubes are to be joined permanently, one may use cyclohexanone, a solvent for polyvinyl surfaces, to make a firm seal (474, 475). This practice has largely been replaced by the use of nipple connectors (674). Pentz (653, 655) described the use of a brass rod to get tight, long-lasting connections between polyethylene and Tygon transmission tubing. Shaw and Duncombe (759) made polythene to glass connections by immersing the ends of the polythene tubing briefly in boiling water to soften them, then pushing the polythene tubing over the

TABLE 2

Materials Adsorbed on Manifold Tubing

Material adsorbed	Recommended tubing	Reference
Barbiturates	Glass	Blackmore et al. (99)
Estrogens	Glass	Strickler et al. (804)
Formazan (from tetrazolium red)	Glass	Avanzini et al. (48)
Iodine	Acidflex	Geiger and Vernot (322)
	Glass	Bomstein et al. (121)
	Glass or polythene	Goodall and Davies (335)
	Glass or Teflon	Bomstein et al. (122)
Mercury	Glass or polyethylene	Ruzicka and Lamm (707)
Metal dithizonates	—[a]	Briscoe et al. (134)
Methanethiol	Glass or polyethylene	Prescott (678, 679)
p-Nitroaniline	Glass	Haesen et al. (360)
Organophosphorus pesticides	Glass or polyethylene	Ott and Gunther (636)
Phenols	Glass	Anstiss et al. (37)
		Fishman et al. (279)
		Friestad et al. (292, 293)
		Horn (414)
		White et al. (674)
Primary amines (e.g., amphetamine)	Glass	Rutter (706)
Quaternary ammonium compounds	Glass	Auerbach and Bartchy (45)
		Linton et al. (542)
Starch-iodine complex	Glass	Powell (672)
Thorium	—[a]	Prall (677)

[a]See text.

glass and securing the joint with thin copper wire. Ganis (303) used Teflon Swagelok fittings (Crawford Fitting Co.) to connect glass to Teflon trans- mission tubing and gives photographs of the connectors.

Techniques for attaching small- and large-bore Teflon tubing to standard glass fittings via nipples or sleeve connectors are described by Christopherson (185).

IV. MANIFOLD CONSTRUCTION

A. Fittings and Techniques

Photographs of "quick disconnect" fittings are given by Kessler (475) and Brown et al. (141). These coupler devices accommodate eight trans- mission tubes and allow quick interchange of manifold and reagent lines. Brown et al. (141) describe a modification of the fitting that helps prevent leaks which may occur under pressure with the units as supplied by Technicon. White et al. (674) provide a table of delay coils and their internal volumes.

Docherty and Skinner (228, 229, 776, 777) used an ultrasonic bath to prevent buildup of a precipitate in an automated method for the analysis of potassium. On the manifold, excess sodium tetraphenylborate reagent was added to the samples, an equivalent amount of the reagent was precipitated by the potassium, and the excess tetraphenylborate was monitored by an ultraviolet spectrophotometer. The precipitate tended to block the mixing coils placed ahead of a Continuous Filter module and eventually plugged up the system. To alleviate this problem, Docherty and Skinner placed the mixing coils in an ultrasonic bath, which was switched on occasionally to dislodge the precipitate. Alternatively, the bath could be operated continuously using a cooling coil for the flow stream to prevent a temperature rise which would have adversely increased the solubility of the potassium tetraphenylborate.

Davis et al. (221) described an intriguing technique which permitted storage of 40 samples or standards in a single coil during the analysis of serum for folate activity. The method required a 20-hr incubation period, after which the microbiological growths from the samples and standards were to be read out turbidimetrically through a Colorimeter at 550 nm. The storage coil was made from 210 ft of 2.8-mm-i.d. polythene tubing and included plastic stopcocks at either end. The samples and substrate were mixed on the manifold and fed sequentially into the coil. After the stopcocks were closed off, the coil was removed from the AutoAnalyzer and incubated 20 hr at 37°C. It was then returned to the AutoAnalyzer for the turbidimetric analysis. With a coil of this size, precautions had to be taken to balance the pressures at the inlet and outlet; otherwise, an unsatisfactory flow through the Colorimeter resulted.

Amador (22) gives instructions on the preparation of polyethylene coils for use with an AutoAnalyzer. Mixing coils, Heating Bath coils, and time-delay coils may be made from polyethylene tubing, 2.15 mm i.d. and 3.25 mm o.d. Each meter of tubing contains about 3.63 ml of liquid. To construct a mixing coil, for example, one selects a test tube of the desired size, e.g., 13 mm, wraps the tubing around the test tube, and secures it with Scotch Tape. The assembly is dipped into boiling water for 1 min, and the tape and test tube are removed, yielding a polyethylene mixing coil. Amador provides photographs of several coils fabricated in this manner and states that such coils may be substituted for glass coils, retaining excellent bubble patterns. They are connected to standard glass fittings via sleeves made of 1.5-cm lengths of Tygon tubing, 0.125 in. i.d. x 0.1875 in. o.d. The economic advantages of easily constructed poly-ethylene coils over fragile glass coils are obvious. Amador mentions two limitations of polyethylene coils: The plastic is destroyed in oil baths above 56°C, and certain antifoam agents tend to coat the coils. Neeley (621) prepared polyethylene coils from 0.062-in.-i.d. tubing by Amador's method, except that the tubing was immersed in boiling water for 2 min.

When operating multichannel analyzers, it is often desirable or necessary to phase the various channels so that the samples arrive at their respective readout devices at the correct points in time. In lengthening a circuit, Gardanier and Spooner (306) caution against the use of mixing coils, recommending instead the use of vertical coils of transmission or glass tubing. Van Belle (858) describes a method for phasing a Technicon SMA 12. Eight U-shaped glass tubes per channel were arranged so that the required number could optionally be added to the flow stream, allowing phasing in time intervals from 5 to 40 sec. Brodie and Hanok (136) designed a wooden upright rack which neatly held the excess transmission tubing required to phase the various channels of a Technicon SMA 12/30.

Arrangements to prevent deterioration of reagents by the atmosphere have been described. The air entering a reagent reservoir as the reagent is pumped out may be pulled through a gas absorption tube containing an appropriate absorbent. Soda-lime (493) or Mallcosorb (Mallinckrodt) (468) has been used to trap carbon dioxide. Lenard et al. (528) used nitrogen to segment manifold flow streams in an automated study of the hydrolysis rates of peptides and amides. The nitrogen, at 0.7 psi, was led from a tank into a bottle about half-filled with ninhydrin reagent. The nitrogen bubbled upward through the reagent and then passed to the nitrogen pump tube inlet. The same bottle served as a source of ninhydrin reagent to the manifold, and the arrangement prevented the oxidation of the reagent by air.

Special equipment has been used to continuously prepare fresh reagents. Shaw and Duncombe (758, 759) show an apparatus for the production of a "constant" or "continuous" culture which was employed in AutoAnalyzer methods of antibiotics analysis. Greely et al. (346) used a similar apparatus in an automatic neomycin assay. Berry and Crossland (78) prepared a diazonium reagent by mixing solutions of 4-nitroaniline and sodium nitrite on a manifold. The amount of the reagent needed to couple with the sample was withdrawn through a modified C4 stream splitter fitting, allowing excess reagent to flow to waste. To modify the C4 fitting, a 0.015-in.-i.d. pump tube was inserted into the lower (exit) limb of the C4 and a 0.090-in.-i.d. pump tube was used as a sleeve fit. The 0.015-in.-i.d. tube was inserted into the C4 chamber so as to withdraw only freshly prepared diazonium salt. A drawing of the arrangement appears in their paper.

Several authors have employed valves to open or close liquid or air lines in continuous flow manifolds. Thomas (837) and Vandermeers et al. (859) describe solenoid pinch valves controlled by timer-cam-microswitch systems and used to open or close lines to the Proportioning Pump as needed. Vandermeers et al. (860) used a solenoid valve to control the flow of air from a Proportioning Pump line. Haines and Anselm (362) designed two dilution loops on a manifold, one yielding a 5-fold dilution, the other a 25-fold dilution. Through electrically controlled two-way valves, four conditions could be selected (no dilution, 5-fold, 25-fold, or 125-fold dilution) before the stream passed to the analytical portion of the manifold. Schroeter and Hamlin (740) used a three-way valve to intermittently send a flowing stream through a filter, as described above in Sec. II, H. Strickler et al. (806) reported an AutoAnalyzer system for the analysis of butanol-soluble and butanol-insoluble iodine in serum and employed a timer and valve system to automatically prepare the butanol extracts. The extracts were sent to a fraction collector for subsequent analysis on various AutoAnalyzer manifolds. Cantor (164) modified the method of Yunghans and Monroe (928) for atmospheric aldehydes by adding a valve system to permit fast, smooth introduction of standards.

Eveleigh et al. (251) show a programmable multichannel valve that pinches or releases tubing via a series of adjustable rollers. There are 12 sets of these rollers and each set is rotated sequentially into contact with 8 tubes. The device can be activated by electrical signals to open or close the tubes so that reagents can be switched or flows can be directed to different portions of a manifold. Bradley and Tappel (126) used such a valve as part of a complex AutoAnalyzer system for the analysis of multi-enzymes.

B. Hints on Building a Manifold

We now consider a few pointers on the physical construction of manifolds. To minimize carryover, one should place glass fittings tightly together, using butt joints with tubing sleeves where possible (185). Angles or pockets in tubing, which permit small portions of the stream to lag behind or which interfere with bubble patterns, should be avoided (248). Glass fittings on the outlet of pump tubes should be connected directly to the pump tubes; intermediate lengths of transmission tubing should not be used (706).

One should keep in mind several concepts when employing very slow pump rates on a manifold. Under such conditions, Crerand et al. (203) noted that reduction of the internal diameters of coils and tubing from 2.0 mm to 1.6 or 1.0 mm improved their performance. Hoober and Bernstein (411) also found that standard mixing coils and Heating Bath coils were not adequate with low pump rates. They made a mixing coil by wrapping a 24-cm length of standard transmission tubing around a glass rod, 0.9 cm in diameter, and a Heating Bath coil by winding a 54-cm length of Solvaflex transmission tubing around a glass tube, 1.0 cm in diameter. The length of the latter coil could be varied to obtain the desired residence time in the Heating Bath. In order to conserve on reagent consumption, Place and Hardy (663) operated an AutoAnalyzer at slow speed and used coils of reduced internal diameter in the manifold. Pollard and Waldron (666) used very slow flows through delay coils to obtain a 3.5-hr treatment in a constant temperature bath. The construction of these coils was critical in avoiding surging. Vertical tubes were not used, and plastic tubes were not used for the connections. Rather, glass to glass butt joint connections were made, in which the end faces of the glass tubes were ground flat with carborundum.

V. SPECIALIZED MANIFOLD DESIGNS

Manifolds have been reported which provide separate chemical paths for sample and blank channels. Bell et al. (71) used two Colorimeters and a dual pen Recorder to simultaneously measure blank and test specimen signals in an automated method for alkaline phosphatase. Kream et al. (498) split a flow stream containing steroids into two portions. One portion was reacted with a phenylhydrazine colorimetric reagent, the other with a "blank" reagent from which phenylhydrazine had been omitted. The two streams were monitored by two Colorimeters, providing sample and blank values automatically. Additional references to this technique can be found in Chapter 3, Sec. VIII, B.

Kabadi et al. (447, 448) used a closed-loop manifold to study the effects of time and temperature on the absorbance of a reaction mixture. The mixture was prepared manually and placed in a reservoir vessel. The Proportioning Pump aspirated the solution from its reservoir and pumped it through a Heating Bath, through a spectrophotometer flowcell, and then back to the reservoir.

Fasce and Rej (256) performed automated kinetic assays of serum lactic dehydrogenase. After the reagents were mixed with the sample, the stream was split into three portions. Each portion was resegmented and each was passed through a separate delay coil. The coils, of different lengths, produced delay times of 0.4, 1.8, and 3.6 min. The streams from the coils were then combined and sent to a fluorometer flowcell. The sampling program was 10 samples per hour, 40 sec of sampling and the remainder wash. Thus, each sample produced three peaks of increasing height with almost complete return to baseline between each peak. The increase in peak height could be correlated to the time-delay values and thereby gave a measure of reaction rate. This method of measurement was felt to be more accurate than the use of a single point for each sample or standard.

Another approach to kinetic measurement of enzyme activity is reported by Vandermeers et al. (860). The sample is mixed with reagents on the manifold, incubated in jacketed mixing coils, and read in a spectrophotometer by a special timer-controlled program. The sample is aspirated for 1.7 min; during this time a solenoid opens an air line from the Proportioning Pump, aiding in pushing the sample completely into the incubation coils. Next, the solenoid closes the extra air pump line, reducing the flow rate of the sample through the incubation coils; simultaneously the Sampler is indexed to its next position, and its power is cut off to hold it there. During the next 3.2 min, the sample flows slowly through the incubation coils and through the remainder of the manifold, eventually reaching the spectrophotometer. The cycle then repeats, processing the next sample. The recorded output appears as a rapid rise at first, changes to a slow, linear increase in signal (the 3.2-min kinetic measurement), and then returns to the baseline or to the next sample. As an additional convenience, the recorder was synchronized so as to be active only during the linear portion of the signal (the kinetic measurement), conserving recorder paper and simplifying the reading of the charts.

Vandermeers et al. (859) have also published a method for the analysis of lipase, trypsin, and chymotrypsin by automatic rate analysis via titration at constant pH. The sample plus substrate were pumped into a jacketed titration cell, where they were mixed by a magnetic stirring bar. A motorized buret (Metrohm Combititrator 3D) and a micro pH electrode were used to maintain constant pH as the enzyme and substrate reacted. The amount of sodium hydroxide consumed gave data useful in the estimation of hydrolases. The apparatus was controlled by a clock-cam system and included a series of solenoid pinch valves to open or close flow lines as needed. The apparatus automatically delivered sample plus substrate to the titration vessel, stopped the Pump and activated the buret-titrant delivery recorder system for the desired time interval, restarted the Pump to flush the spent sample from the titration cell and provide several rinses while the buret simultaneously refilled, brought fresh sample and substrate to the titration vessel, and commenced a new cycle.

Continuous flow manifolds may be designed for the determination of acids or bases through the addition of a buffer-forming reagent (usually a weak acid or weak base), followed by the addition of an indicator dye solution and subsequent readout of the stream transmittance in a color-imeter. Cool and Annokkee (197) give a theoretical discussion of such systems, including the required mathematics and a procedure for selecting the proper analytical wavelength. Graphical treatment of the equations allows selection of reagent and sample strengths which will yield good linearity over wide ranges of transmittance. Data are presented from eight different analytical methods for acids and bases, including sodium barbiturate. The standard deviation of the measured percent transmittance from linearity was typically within 0.3 to 1.0, i.e., linearity was usually excellent. Cool and Annokkee recommend holding the temperature within $1^{\circ}C$ when using such manifolds, since the dissociation of weak acids and weak bases is temperature dependent. These authors also cover the mathematics needed to predict the theoretical accuracy and precision that may be expected from analytical systems of this type. A special set of equations is required for these estimates, because Beer's law as such does not hold for buffer-indicator acid-base determinations.

Chapter 5

THE USE OF CONTINUOUS FLOW MANIFOLDS
WITH LIQUID CHROMATOGRAPHIC SYSTEMS

I. INTRODUCTION

A complete discussion of automated liquid chromatography would lie far beyond the scope of this book. Rather, we shall touch on the various uses of continuous flow manifolds with column chromatographic systems. Such uses may be numbered among the earliest specialized applications of the continuous flow technique (264, 771). Jones (442) has written a valuable review of automated peptide chromatography, in which uses of the Technicon Amino Acid Analyzer and AutoAnalyzer systems are discussed.

The commercial availability of high-pressure liquid chromatography (HPLC) instruments now renders almost routine many separations considered nearly impossible a few years ago. Nevertheless, most currently marketed HPLC systems feature detector modules restricted to measurements that depend on intrinsic properties of the solutes, i.e., their ability to absorb energy in the uv-visible portion of the spectrum, to fluoresce, or to influence the refractive index of the solvent-solute mixture. It therefore remains of interest to review methods of effecting physicochemical transformations, using the continuous flow technique, which permit either general or specific detection of the solutes emerging from a liquid chromatographic column.

Several flexible arrangements are available to the experimenter. One may control the flow of the eluant through the column with a Proportioning Pump or a separate high-pressure pump. In the off-line mode, portions of the column eluate may be obtained with any suitable fraction collector and may be analyzed later by aspirating the fractions into a continuous flow manifold. In the on-line mode, the column eluate may be aspirated directly into the manifold, or it may be split into two streams, one flowing to the manifold, the other to a fraction collector. Examples of all these configurations may be found in the references listed in Table 1. The listing is presented to enable an experimenter to locate papers in his or her area of interest, and it should not be considered as comprehensive.

Perhaps the first decision that must be made is whether one will use an off-line or on-line arrangement. Schroeder (737) points out that the off-line mode allows a single AutoAnalyzer to service the output of several columns, whereas the on-line mode requires a separate manifold, detector, and recorder channel for each column. Manual treatment of column fractions before presenting them to the AutoAnalyzer may be desirable. Thus, Schroeder and Robberson (736) collected fractions of peptide samples from ion exchange columns and manually hydrolyzed the fractions before analyzing them on a ninhydrin colorimetric manifold.

On the other hand, on-line operation permits observation of finer chromatographic separations without the loss of detail caused by the averaging effect of fractions. If the continuous flow analytical system is sufficiently sensitive, it may require only a small fraction of the column eluate to record the chromatogram, allowing the excess eluate to be diverted to a separate fraction collector for subsequent additional tests. If these additional tests are tedious or expensive, the on-line chromatogram may prove very helpful in selecting only those fractions of interest for further study.

Continuous flow operation ahead of the column itself may also be useful. For example, Grasbeck and Karlsson (341) used a Proportioning Pump to mix 3 $\underline{N}$ sodium hydroxide with a protein sample and then fed the mixture to the column inlet.

II. OFF-LINE APPLICATIONS

Schroeder et al. (735) collected fractions of peptide samples from ion exchange columns. The fractions were aspirated into an AutoAnalyzer manifold from a Sampler I module. Special test tubes that would fit in the Sampler tray and, with adapters, into the fraction collector were employed, thus reducing sample transfer time. Jenner (437) described experiments to measure the molecular weight distribution of dextran. The eluates from two columns were collected into Technicon sample cups using a fraction collector that serviced both columns simultaneously. After the fractions had been delivered into concentric rings of cups, the sample tray holding

TABLE 1

Continuous Flow Manifolds Used with Column
Chromatographic Separation Systems

Material studied	Mode[a]	Reference
Alcohols	On-line	Santacana and Mitzner (718)
Amino acids	Off-line	Cunningham et al. (212)
		Taylor and Marsh (824)
	On-line	Barber (62)
		Blaedel and Todd (111)
		Block and Mandl (116)
		Brown (144)
		Buist and Strandholm (151)
		Cadavid and Paladini (158)
		Ertingshausen et al. (245, 246)
		Ferrari and MacDuff (272)
		Stegink (790)
		Thomas et al. (836)
		Thompson (839)
		Thompson et al. (840)
Amino sugars	On-line	Gregory and Van Lenten (348)
Amylase	Off-line	Wilding (910)
Carboxylic acids	On-line	Zerfing and Veening (938, 939)
Catecholamines	Off-line	Merrills (593)
	On-line	Merrills (594)

TABLE 1 (continued)

Material studied	Mode[a]	Reference
Chemical oxygen demand	On-line	Zuckerman and Molof (941)
Cystine, cystinyl peptides	On-line	Walsh et al. (889)
Dextran	Off-line	Jenner (437)
Enzymes	Off-line	Beck and Tappel (68)
	On-line	Catravas and Lash (174)
		Chersi et al. (182)
		Hoober and Bernstein (411)
		Roodyn (692)
	Both	Tappel and Beck (822)
Estrogens	On-line	Strickler et al. (805)
Glycopeptides	Off-line	Johnson et al. (439)
Glycopeptides, glycoproteins	On-line	Brummel et al. (148)
Guanidinosuccinic acid	On-line	Kamoun et al. (451)
Hexosamines	On-line	Lee et al. (527)
Hexoses	On-line	Weber et al. (895)
5-Hydroxyindoleacetic acid	On-line	Brown et al. (146)
Lactic dehydrogenase	Off-line	Hochella and Weinhouse (405)
Levan (in dental plaque)	On-line	Manly et al. (566)
Lysine (derivatives)	On-line	Rexen and Christensen (685)
Neomycins	On-line	Kaptionak et al. (452)
Nucleotides, acid soluble	On-line	Weinstein et al. (896)

TABLE 1 (continued)

Material studied	Mode[a]	Reference
Peptides	Off-line	Schroeder (737)
		Schroeder et al. (735)
		Schroeder and Robberson (736)
	On-line	Armstrong (43)
		Buist and Strandholm (151)
		Catravas (172)
		Delaney (223)
		Easley et al. (235)
		Hill and Delaney (397)
		Jones (441, 442)
		Lindley and Haylett (538)
Phosphates	On-line	Benz and Kelley (73)
		Benz and Paixao (74)
		Czech and Hrycyshyn (213)
		Heinke and Behmann (383)
		Lundgren (554)
		Lundgren and Loeb (555)
		Pollard et al. (669)
Phosphate ester metabolites	On-line	Jellinek et al. (436)
Phosphorus-containing compounds	Both	Mandl et al. (563)
Proteins	On-line	Bartley and Poulik (64)
		Chersi et al. (182)
		Mandl (562)
		McKenzie et al. (587, 588)
		Roubal and Tappel (697)

TABLE 1 (continued)

Material studied	Mode[a]	Reference
Ribonuclease	On-line	Mundry (616)
Ribonucleic acid	Off-line	Goldstein et al. (334)
Steroid phosphates	On-line	Beyer and Morozowich (88)
Sugars	On-line	Catravas (173)
		Etter (247)
		Kesler (471, 472)
		Larsson and Samuelson (521, 522)
		Lee et al. (526)
		Samuelson et al. (713)
		Samuelson and Swenson (714)
		Wood and Cousins (917)
Sugar phosphates	Both	Burt (156)
Thyroxine	Off-line	Kenny et al. (469)
		Webb (892)
Ureas	On-line	Santacana and Mitzner (718)
Xanthurenic acid	Both	Looye et al. (548)

[a]Off-line: eluate fractions collected, then sampled and analyzed on manifold. On-line: manifold continuously monitored column eluate.

the two series of fractions was placed on a specially constructed sampler module. The samples were processed through two duplicate manifolds on one Proportioning Pump and a double Colorimeter fed the results to two Recorders. The system doubled the normal output of chromatograms.

The Technicon Sampler II may be used as a fraction collector. Beck and Tappel (68) collected 7.7-ml fractions into cups on a Sampler II equipped with an ice bath. The fractions were then tested for various enzymes using an AutoAnalyzer system. Tappel and Beck (822) published an interesting example of a system using both off-line and on-line modes. Enzymes eluted from an ion exchange column were monitored on-line by passing the column eluate through a fluorometer flowcell. After exiting the flowcell, the stream was collected in 7.7-ml fractions in a Sampler II equipped with a cooling bath. The same Sampler II was used to feed samples of each fraction to a separate manifold for additional colorimetric analyses using various substrates. This operation is described in more detail in Chapter 2, Sec. X,D,2, to which the reader is referred.

Merrills (594) studied radioactively labeled tyrosine and its metabolic products. The column eluate was split into two portions. One stream was continuously monitored by an AutoAnalyzer system and the other was sent to a fraction collector for later analysis in a radioactivity counter. Goldstein et al. (334) used an AutoAnalyzer manifold to perform preparative chemistry on fractions from column eluates. The fractions were sampled by a Sampler II and, after passing through the manifold, they were automatically deposited by a specially built probe mechanism onto paper disks that were held in a custom-built wheel. The sample deposition system was synchronized to the AutoAnalyzer by a signal from a Colorimeter. The Colorimeter was arranged to monitor the stream for a dye solution continuously pumped to the Sampler II wash box. A rise in transmittance indicated that the wash solution was leaving the Colorimeter flowcell and that a sample was appearing in the stream. This rise in transmittance triggered the timed deposition of sample from the flow stream onto the paper disk in the output sample collection device. The disks were subsequently placed in a scintillation counter, and its readings allowed the plotting of chromatograms of the radioactive materials.

III. ON-LINE APPLICATIONS

When an AutoAnalyzer is used for on-line monitoring of column eluates, it is especially important that the manifold respond rapidly to changes in concentration of the incoming stream (248). A slow rise time in a system used to monitor a very efficient column can cause an apparent loss in resolution of closely eluting bands (836). All available techniques for reducing carryover must be brought to bear when constructing a manifold to produce the recording of on-line chromatograms. Mandl (562) constricted the outlet of a column tube to a 2-mm bore to prevent loss of resolution due to mixing. Ertingshausen et al. (245, 246) have published a particularly useful account of how various AutoAnalyzer components were modified to reduce carryover and thus to improve the fidelity of the recorded chromatogram. These authors also discuss the theory of sample spreading in the AutoAnalyzer. Their paper deserves careful study if an experimenter wishes to design an AutoAnalyzer system capable of approaching an actual observation of true column chromatographic resolution.

Multiple manifolds, each monitoring a portion of the column eluate, can yield valuable complementary data from a single chromatographic run. Weinstein et al. (896) separated acid-soluble nucleotides on an anion exchange column. A portion of the eluate was passed to an ultraviolet spectrophotometer, and another portion was sent through an AutoAnalyzer manifold to record the phosphorus content. Steginck (790) used radioactively tagged amino acids in studies that converted some of the compounds to ninhydrin-inactive metabolites. The column eluate was split between an AutoAnalyzer manifold and a flowcell placed in a scintillation counter. Catravas and Lash (174) chromatographed enzymes and used three Auto-Analyzer channels to simultaneously monitor the column eluate for alkaline phosphatase, glutamic oxalacetic transaminase, and total protein. Chersi et al. (182) separated esterases on a Sephadex column and monitored the column eluate for phosphatase, phosphodiesterase, and protein using three channels on an AutoAnalyzer manifold. Barber (62) used multiple manifolds to monitor column eluates in studies of sulfur amino acids. The different chemistries helped make identification of the eluting bands more positive. In some cases, unresolved (simultaneously eluted) pairs of compounds could be analyzed by the relative responses of two or more analytical systems.

Another approach to multiple testing of column eluates was described by Roodyn (692). A Sampler module was arranged to sequentially deliver various substrates as reagents to the AutoAnalyzer manifold as it monitored the eluate stream. Repeating sequences of substrates were placed around the Sampler tray (ten substrates in four similar groups), and the system conducted ten different enzyme determinations per milliliter of column eluate.

Continuous flow monitoring is also of value in selecting and optimizing column chromatographic parameters. Bartley and Poulik (64) used an AutoAnalyzer to help evaluate chromatographic systems for protein isolation. A miniature column was employed, and each drop of eluate was individually analyzed by the AutoAnalyzer. Instead of a direct connection between the column outlet and the AutoAnalyzer, a funnel was placed below the column so that a drop-air bubble-drop-air bubble sequence passed to the inlet of the Proportioning Pump. The recorded trace thus consisted of a series of very sharp, closely spaced peaks whose tops could easily be blended together by visual inspection into a chromatogram.

Once a chromatographic system has been selected, an AutoAnalyzer may be useful in checking lot to lot variation in the chromatographic material. Looye et al. (548) routinely analyzed xanthurenic acid in urine samples by passing the samples through Sephadex G-10 columns and then assaying the purified sample fractions on an AutoAnalyzer in the off-line mode. New batches of the Sephadex G-10 were standardized by following the elution with an on-line AutoAnalyzer to locate the correct fraction.

Santacana and Mitzner (718) employed an AutoAnalyzer not only to monitor an ion exchange resin column eluate, but to regenerate one column while the other was in use.

Ertingshausen et al. (245, 246) mention several fail-safe devices used with a high-pressure column chromatograph (Technicon Amino Acid Analyzer). The Pump III was equipped with a leak detector. Pressure relief valves on the column lines were set at 950 psi. If the valves became activated, the resulting pressure drop caused the flow to reverse in the flowcell, pulling air into it. The large change in photometer signal then triggered a shutdown circuit, which was also activated if the photometer lamp burned out.

IV. SPECIAL APPLICATIONS

Improved separations and shortened run times are two advantages which may accrue from using gradient elution chromatography.* The gradient eluant may be formed by a separate, commercially available device such as the Technicon Autograd (173, 235) or by using a Proportioning Pump to deliver a strong solution to a mixing flask while withdrawing the resulting gradient mixture and pumping it to the column head or to a separate column pump (213, 383, 594, 718, 938, 939). Further discussion of the latter technique is given in Chapter 4, Sec. II, L.

Beyer and Morozowich (88)[§] give the following equation for calculating the concentration of an electrolyte in a gradient eluant:

$$-\ln \frac{c_i - c_t}{c_i} = \frac{rt}{v} \tag{1}$$

---

*For additional examples of gradient elution chromatography, see Refs. 68, 73, 554, 563, 822, and 895.

[§]In the original paper, Eq. (1) is given in the form $-\log\left[(c_i - c_t)/c_i\right] = rt/v$. The log is to be taken on base e (private communication from William F. Beyer, June 3, 1974).

where c_i = concentration of electrolyte in feed reservoir

c_t = concentration of electrolyte in mixing vessel at time t

r = flow rate, ml/min

t = time, min

v = volume of solution in mixing vessel

The flow rate r at which the electrolyte stock solution is moved into the mixing vessel equals the flow rate at which the gradient eluant is removed. Equation (1) is another form of Eq. (1), Chapter 4, and is more thoroughly discussed in Chapter 4, Sec. II, L, 1. Benz and Kelley (73) also cover the mathematics required to calculate the strength of a gradient eluant at the point of elution of a chromatographic band.

Continuous flow polarography has been used to monitor chromatographic column eluates. Blaedel and Todd (111) reacted α-amino acids in a column eluate by passing the stream through a tube containing 50 to 60 mesh copper phosphate. An equivalent amount of Cu(II) was liberated and monitored by their continuous flowcell polarographic system (110). Buchanan and Bacon (150) monitored ion exchange columns via continuous square-wave polarography.

Pumps other than those supplied by Technicon have been used in continuous flow chromatographic systems. Kamoun et al. (451) used an Ismatec MP 13 peristaltic pump to pump samples through an ion exchange column and the eluate through a colorimetric manifold. Vestergaarde and Vedso (872) employed a Beckman Model 746 solution metering pump, featuring Teflon valves, which would handle solutions that destroyed pump tubes ordinarily used with a Technicon Proportioning Pump.

Vestergaarde and Vedso (872) also described a special apparatus designed to process fractions collected from a multiple-column chromatography system. The columns were serviced by a fraction collector. After the colors in the fractions were manually developed, the solutions were fed through a Technicon Colorimeter using a timer system to control the Beckman 746 pump and the fraction collector. A special sampling device, also controlled by the timer system, acted as a probe to raise and lower the sample line into the tubes as the fraction collector indexed.

Lundgren and Loeb (555) give the mathematical treatment of elution curves obtained in the separation of orthophosphate, pyrophosphate, tripolyphosphate, and other phosphates on an ion exchange resin column. Pollard et al. (669) discuss the mathematics required to obtain the areas under chromatographic peaks in terms of absorbance when the recorded charts are linear in percent transmittance.

Chapter 6

AUTOMATIC DATA PROCESSING
IN CONTINUOUS FLOW ANALYSIS

I. SYSTEMS OPERATED WITHOUT DIGITAL COMPUTERS

A. Data Acquisition Systems

Owen et al. (640) describe an analog-digital printout system used to acquire data from AutoAnalyzers. A precision potentiometer was placed on the pen drive motor of the Recorder. The signal from the potentiometer was modified to conform to the AutoAnalyzer calibration curve. A remote data readout device converted the corrected signal to a digital format, and the results were printed on an electric typewriter. Farr et al. (255) reported a somewhat similar approach used to obtain concentration data from a four-channel AutoAnalyzer system. A retransmitting potentiometer, placed on the Technicon Recorder, sent an analog voltage to a peak hold circuit. The voltage measured at the peak maximum was converted through a diode function generator to a voltage numerically equal to concentration. As the voltage dropped away from the peak maximum, a digital voltmeter was triggered to read out the concentration into an electric typewriter.

Sawyer et al. (727) designed a printout system around a modified Technicon Recorder to aid in the analysis of the "original gravity" of beer. The AutoAnalyzer was comprised of two channels, one measuring alcohol, the other sugars, and the signals from the two channels were sent to a dual pen Recorder. The two results, after some calculations and subsequent addition, gave a measure of the original gravity of the beer. To automate the readings and calculations, attachments were placed on both of the Recorder pen drive shafts. On each, a slipping clutch was set to drive a lever and was arranged to trip a microswitch as the Recorder shaft changed directions, that is, as it fell away from a peak. The microswitch triggered a digital converter which printed the value on an adding machine tape. The digital converter received its signal from a 1000 parts per turn digitizer. This digitizer was not connected directly to the Recorder pen drive shaft, but rather was driven by a cam whose periphery was pulled around by a wire attached to the Recorder pen drive shaft. The cam shape was calculated from calibration curves obtained from the associated AutoAnalyzer channel. Thus, the Recorder shaft rotated as a function of percent transmittance but, through the action of the cam, the digitizer rotated as a function of the alcohol (or sugars) content. The calculations, therefore, were done mechanically. The two AutoAnalyzer channels were phased so that, first, a peak from channel 1 was detected and the result printed, and then a peak from channel 2 was detected and its result printed. Addition of the two values was triggered by the main control system, giving the final desired result, and the system was cleared for the next peak from channel 1. Sawyer et al. (727) give details on the main control system and on how the cam shapes were calculated.

Hellerstein et al. (385) used a Gilford Model 300-N photometer with a Model 4006 Data Lister to monitor an AutoAnalyzer. The Data Lister was provided with a switch manually depressed to activate the data printer. To automate the data acquisition step, a synchronous timer was arranged to trip the switch at regular intervals. The timer was set to print a reading every 30 sec on the baseline and was then set to print a reading every 2 sec when the peaks began coming out. The printer tape was scanned visually to locate the maximum absorbances representing the value of each peak.

Dalal and Winsten (214) used a T and T Technology Incorporated Digital Concentration Analyzer, Model DC 801-Dual, with several Auto-Analyzer systems. In one determination (chloride), the standard curve did not pass through zero, which was one of the requirements for direct digital concentration readout with this instrument. The problem was solved by simply adding the required value (60 in this instance) to the printed values. Thus, a value of 50 on the Victor printer was converted to 110 mEq chloride per liter.

B. Programmable Calculators

Lines (540) discusses the use of an Olivetti Programma 101 for calculation of AutoAnalyzer results in clinical laboratories, including the storage of daily standard curve data on resuable magnetic cards, and a method to account for drift. Herman et al. (391) describe a program for this calculator, including an approximation for curvilinear systems wherein the AutoAnalyzer absorbance values are not exactly linear with concentration. The complete program is given in their paper.

Hormann et al. (413) used an AutoAnalyzer to prepare soil samples of herbicides for subsequent analysis by gas chromatography (see Chapter 3, Sec. V). They describe a program to calculate results from the gas chromatograph using a Wang 700 calculator system.

II. SYSTEMS OPERATED WITH DIGITAL COMPUTERS

A. General Descriptions of Computer Systems

Digital computers have proven highly useful not only in acquiring and processing data from AutoAnalyzers, but in laboratory management and control as well. The papers discussed in this section will be of use in getting a bird's-eye view of computer operations with AutoAnalyzers and other continuous flow analyzers.

Much informative and practical research on computerization has been carried out in clinical laboratories. Laessig et al. (515) described a data processor for the Technicon SMA 12/60 which yielded results in digital form. Punched paper tape could be produced for off-line computer processing, if desired. Poletti et al. (665) discussed how samples were identified, controlled, and computer processed from an SMA 12 in a hospital. Hicks et al. (393) used a LINC computer fitted with 8 K of core,* two tape drives, and one disk drive, along with LABCOM programs to monitor up to 24 instruments producing slow analog signals, such as SMA 12s, AutoAnalyzers, or atomic absorption spectrometers. Brecher and Loken (131) described and evaluated a PDP-8L computer connected to eight AutoAnalyzers. The computer, disk and disk controller, input/output unit, and card reader were installed in duplicate to permit rapid switchover in case of a failure. Brecher and Loken included a good discussion of sample flow, use of hardware and software, and failure rates. Van Dreal et al. (861) operated an IBM 1800 computer (24 K of core, two disk drives) with five AutoAnalyzers and an SMA 12/30 on-line. In addition, a second hospital fed data from seven AutoAnalyzers and an SMA 12/30 to the computer via an analog-to-digital converter and a phone line link. In the book Automation and Data Processing in the Clinical Laboratory (47),

---

*Eight K of core means a central memory of 8 x 1024 words. A "word" is one storage location, roughly speaking.

several hospital laboratory systems including AutoAnalyzers interfaced to computers are described. Also discussed are various hospital management computer systems, some successful, others not. Gould and Brooks (336, 337) described the Technicon computer system from the viewpoint of hospital control, records management, and the monitoring of AutoAnalyzer and SMA 12 systems.

When limited numbers of AutoAnalyzers are being operated, it is usually uneconomical to directly interface the systems to a large on-line digital computer. Several laboratories prefer off-line data collection for subsequent batch processing. Sparagana et al. (784, 785) used an analog-to-digital converter driving an IBM card punch to collect data from an AutoAnalyzer for off-line calculations and published the FORTRAN program used with an IBM 1620 computer (785). Constandse (194, 195) described the IBM 1080 Data Acquisition System and its use with AutoAnalyzers. Eckman et al. (236) used the IBM 1080 system to convert the AutoAnalyzer analog signal to digital form and to drive an IBM card punch. The punched card data were then processed off-line by a computer.

A small computer may be used to collect and process AutoAnalyzer data for subsequent transfer to a large computer. Stewart et al. (798) used a LINC-8 computer for data acquisition, calculation, and report writing. The test results were then transmitted to the control computer, an IBM 360, via punched tape or punched cards.

Several interesting general discussions of computer programs have appeared. Torud (846) contrasts three computer programs applied to typical AutoAnalyzer curves. The errors, precision, advantages, and disadvantages of each are discussed and compared to the use of the Technicon Chart Reader. Alsos et al. (20, 21) published a flow chart of a FORTRAN IV program applied to AutoAnalyzer curves. Although not detailed, this flow chart should be useful to someone starting to write such a program. Hosley et al. (415) give a program in BASIC for an Auto-Analyzer time-shared computer system. Evans and Thomas (248) discuss computer processing as applied to automated analysis in nutritional research.

B. Interfaces

1. Mechanical Data Transducers

The two most commonly employed mechanical data transducers are retransmitting potentiometers and shaft encoders. A retransmitting potentiometer is simply a high-quality variable resistance. Almost invariably, it is attached directly to the main drive shaft of the Technicon Recorder. By applying a constant input voltage across the potentiometer, the output voltage becomes an analog representation of the pen position. Since the signal is in analog form, an additional device must be used to obtain the signal in digital form. A shaft encoder differs from a retransmitting potentiometer in that a signal in digital form is immediately available from the transducer.

Digital shaft encoders on AutoAnalyzer Recorders were used by Young (927) to produce data on punched paper tape. The tapes were processed off-line on an IBM 1130 computer. Kelly and Wilson (463) also interfaced an AutoAnalyzer to an off-line computer by placing a shaft angle encoder on the Recorder. The Recorder shaft position was stored by a control chassis which converted the angle to binary coded decimal (BCD) form, and these data were punched on paper tape every 7.5 sec by a Datex system. Identification and dilution data could also be punched on the tape. The computer read the tape, picked the peaks, performed the calculations, and produced the report.

Retransmitting potentiometers are more widely used than shaft encoders. There are several ways to treat the analog voltage after it leaves the transducer. Perhaps the most straightforward approach, at least for off-line data collection, is to monitor the voltage with a digital voltmeter, feeding the digitized voltage data to a paper tape punch. Abdullah (1) employed such a system and incorporated a scanner which permitted data collection from several Recorders. Timers were used to activate the data collection system. Cheek (180) used a retransmitting potentiometer on an AutoAnalyzer Recorder to feed analog signals to a multiplexer. After sequential sampling of the signals by the multiplexer, the signals were sent through a digitizer, and the data were recorded on tape for later batch processing on a computer.

Greenway (347) described a peak sensing system for data collection. A retransmitting potentiometer sent the analog signal to a storage capacitor. The stored voltage was compared to the actual (incoming) voltage and, when a peak dropoff was sensed, a logic pulse triggered a voltage-to-frequency converter to send the stored value through a digitizer to a card punch. Similarly, Caisey and Riordan (160) fitted AutoAnalyzer Recorders with retransmitting potentiometers and fed the signals to a peak detector device. The schematic for this circuit is given in their paper. The detector stored the incoming voltage and continuously compared the stored value against the incoming value. Thus, as the Recorder moved upscale approaching a peak, the stored and incoming voltages remained equal. As the Recorder dropped away from the peak, the detector stored the highest voltage, corresponding to the peak value, and began comparing this peak voltage to the decreasing incoming voltage. When the difference between the two voltages became great enough, the detector tripped a digitizer-command relay which triggered the recording of the data by a

Compact-2 Datalogger. This system contained a digital voltmeter, a BCD converter, and outputs to a strip printer and a tape punch unit. The data were processed off-line through a phone line link to a remote computer.

Flynn et al. (285) used an AutoAnalyzer interface to recognize each peak, quantitate the peak heights, convert them to binary form, and record the data on punched paper tape. The system handled up to eight Auto-Analyzers. The peaks were sensed through the action of a clutch arm placed on each Recorder shaft and arranged to act on two microswitches mounted close together. The sequential closing of one switch and then the other indicated a reversal of Recorder direction corresponding to either a peak or a valley. The distance between the two switches was adjusted to ignore noise. The "backlash" thus introduced meant that each punched peak value was lower by a constant amount than the actual peak value, and this error was compensated for in the computer program. Event markers were employed to indicate on the recordings when each peak reading was taken, allowing visual checks of the system's performance. In addition, a timer forced a reading to be taken if no peak was sensed within the proper interval. Without this safety precaution, the sequencing of readings could have gone out of synchronization with the order of cups placed on the Sampler. Flynn et al. also discuss their arrangements of punching codes, including methods to code which AutoAnalyzer channel had been read, whether a datum resulted from the peak sensing device or from the timer, and so forth. The value of such peak sensing systems lies in the reduced programming effort required at the computer site. Since the peaks have already been picked, the computer need not be programmed to perform this task.

A further refinement consists of conditioning the data before they are recorded. Collen et al. (190) describe the use of linearization and calibration networks placed between Colorimeters and Recorders. The corrected signals were traced out by the Recorders, and retransmitting potentiometers sent the analog signals to four storage capacitors. The capacitors (representing four Colorimeter channels) were read synchronously by a digital voltmeter which sent the data to an IBM card punch for off-line batch processing.

Retransmitting potentiometers are also useful in on-line computer data acquisition. Evenson et al. (252) used a potentiometer on an Auto-Analyzer Recorder shaft to transmit analog signals to a LINC computer. Werner (897) sent the signal from a retransmitting potentiometer to an analog-to-digital converter and thereafter to a computer. From a standard 1500 Ω retransmitting potentiometer placed on an AutoAnalyzer Recorder, Gray and Owen (345) obtained analog signals for use with a Digital Equipment Corporation PDP 8/S computer. Their interface handled five AutoAnalyzer channels and included indicator lamps showing which

Recorders were on-line to the computer. Carroll (167) interfaced an AutoAnalyzer Recorder to a computer via a recorder-repeating slidewire which fed 0 to 10 V dc to a peak storage device. The computer sampled the peak values from the storage unit on a timed basis. Brown (142) described an AutoAnalyzer-computer interface marketed by Berkeley Scientific Laboratories, which used a retransmitting potentiometer on the Recorder. Alber et al. (14) arranged a Beckman DK2-A spectrophotometer to monitor the effluent from an AutoAnalyzer manifold at a constant wavelength and mounted a retransmitting potentiometer on the pen pulley to send analog voltage signals to a Digital Equipment Corporation PDP-12A computer.

The advantages and disadvantages of mechanical data transducers may be briefly summarized as follows: Wherever a mechanical motion related to the sought data exists, a retransmitting potentiometer may be fitted to obtain the required analog signal; an easily accessible voltage or other electrical signal indigenous to the instrument is therefore not a prerequisite. The electronics involved are quite simple; all that is required is a constant voltage power supply and perhaps a trim potentiometer or two to adjust or calibrate the resistance of the retransmitting potentiometer so as to achieve exactly the desired zero and full-scale voltages. Since the voltage range is determined by the experimenter and not by whatever voltage range happens to exist in the instrument to be interfaced, widely differing instruments can be easily fitted to supply the same output voltage range to the computer input device. When a retransmitting potentiometer is placed on a recorder, the inertia of the recorder pen servo system helps to damp out unwanted high-frequency noise, a distinct advantage when monitoring slowly changing signals. Unfortunately, this very fact renders retransmitting potentiometer-recorder systems unsuitable for following rapidly changing signals. In addition, the moving parts of the retransmitting potentiometer are subject to wear, and they may introduce noise if contaminated with dust or grit. Finally, one must ensure that the retransmitting potentiometer is operated over most of its available range. If the mechanical motion of the instrument turns the potentiometer shaft through only a small portion of its rotation span, large errors due to the resolution limit of the potentiometer may result.

2. Electrical Data Transducers

The sophistication of electrical transducer systems is increasing by leaps and bounds, and today's systems are rapidly becoming obsolete. At this writing, there are two commonly used approaches to electrical data transduction. First, one may tap a voltage present in the instrument to be interfaced which accurately reflects the parameter to be monitored. This voltage may then be amplified, if needed, to suit the requirements of the computer analog-to-digital converter. Second, if the instrument possesses a BCD output (many of the latest designs include such an output as a standard feature) and if a compatible BCD input exists on the computer itself, direct interfacing becomes simply a matter of running the needed connecting cables.

Cotlove (199) obtained computer input signals from the photometer outputs of continuous flow analyzers. Griffiths and Carter (349) tapped the photocell signals on AutoAnalyzer systems and routed the signals through bench consoles, each of which would handle up to eight AutoAnalyzer channels, and thereafter to the computer room via cables. There, the measuring and reference voltages were amplified and sent through a logic switchboard to a single analog-to-digital converter which transferred the digital data to an Elliott 903C computer system. The AutoAnalyzer Recorders were retained to permit visual monitoring of the analytical systems in the laboratory. Stewart et al. (798) used a Digital Equipment Corporation interface to track AutoAnalyzer voltage signals and send them to a LINC-8 computer.

Alber et al. (14) give the schematic of an operational amplifier circuit designed to act as an on-line interface of a Technicon Fluorometer II to a Digital Equipment Corporation PDP-12A computer.

Keay (458) described an arrangement for off-line or on-line interfacing of Precision Electronics (Australia) colorimeters used with AutoAnalyzer manifolds. Each colorimeter fed its signal to its own built-in peak sensor circuit. The peak sensor output from each colorimeter presented its stored peak voltage to a scanner which sequentially conveyed each voltage to an analog-to-digital converter. The digital BCD data were recorded on punched paper tape or were sent directly to a computer. The scanner and punch units were part of a Solartron Compact Data Logger (Australia) system.

Griffiths and Carter (349) discussed an unusual interfacing problem which arose while they were attempting to connect the photocell outputs from a Technicon Flame Photometer to an Elliott 903C computer. The recorded working range of the Flame Photometer lay in the upper 40% of its full-scale output. As supplied by Technicon, through zero offset and a 2.5x scale expansion, the working range was presented as a full-scale signal to the Recorder. When the instrument's photocells were connected to the computer's analog-to-digital converter, the system became grounded, causing the 2.5x scale expansion to be lost on the Recorder. To resolve this situation, they designed a special interface which retained the expanded recording but allowed the computer to read the unexpanded signal. The schematic of the interface is given in their paper.

In considering electrical interfacing with operational amplifier networks, several pros and cons should be kept in mind. On the plus side, operational amplifier circuits afford signal conditioning flexibility not to be found with retransmitting potentiometers. A wide band pass (high-frequency response) may be selected to permit accurate tracking of rapidly changing signals, or a low band pass (low-frequency response) may be selected to eliminate unwanted high-frequency noise and to pass only the desired slowly changing analog signals to the computer. No moving parts are involved, and the service life should be unaffected by dirt or dust in the laboratory environment. On the other hand, individual circuits must usually be designed to handle the various available voltage levels, noise characteristics, or impedances of each different type of instrument to be interfaced. Although many instruments can be conveniently tapped at the recorder terminals, a few present real headaches in trying to locate a suitable analog voltage. Several of the articles cited in Chapter 2, Sec. I, F, 1, discuss circuits which may be useful when electrically interfacing Technicon Colorimeters to a computer. The linearity (accuracy) of operational amplifier circuits is a function of the quality of the amplifiers and of the care used in the interface circuit design. Because of slow amplifier drift, frequent calibration of such interfaces may be manditory.

A direct BCD interface (from the BCD output of an instrument to the BCD input of a computer) affords exact signal tracking capability. Because no analog amplification is required, errors from amplifier nonlinearity or drift are eliminated. Such direct BCD interfaces, however, are very inflexible, since no signal conditioning in the interface itself is possible. Problems arising from instrument noise or "jitter" must therefore by dealt with by the computer programmer.

3. Sampler Interfaces

Interfaces between sampling modules and computers usually serve one of two purposes: (a) sample specimen identification and (b) control or measurement of sampler module actions.

The identities of samples were machine read from precoded sample carriers by assay sampler units in a system described by Cotlove (199). This method of sample identification permitted random sample input and the placement of emergency samples on the sampler module during the run. Constandse (194, 195) discussed an IBM card reader system operated with a Technicon Sampler 40 for positive specimen identification. Gould and Brooks (336, 337) outlined the Technicon computer system which also made use of a card-reading Sampler, the Model T-40.

Computer control of a Sampler formed an integral part of a 10-channel computerized AutoAnalyzer system described by Kirschman et al. (483). The computer, rather than the AutoAnalyzer operator, started the Sampler II module at the beginning of the run. By so doing, the analytical system was prevented from starting before the computer was ready to accept data. In addition, this arrangement allowed the computer to measure the delay time from the aspiration of a sample to the appearance of the corresponding peak, and this information along with peak and valley detection permitted the computer to check on the validity of each peak. Bennet et al. (72) used a program and an interface (consisting of a driver and relay) in place of the usual cam-microswitch system to control the probe action of a Sampler II. The signal times were accurate to 0.02 sec. Taking into consideration the timing errors associated with the interface relay action and those associated with the Sampler II probe arm mechanism, overall timing accuracy was better than 0.5 sec. Gray and Owen (345) located the single-pole microswitch inside the Sampler II that stopped the probe mechanism in the sampling position and replaced it with a two-pole, two-way microswitch. One pole of the new switch acted in place of the original switch, and the other notified the computer when sampling had started or finished.

4. Operator Control Stations

Evenson et al. (252) describe a switch box system that allowed the AutoAnalyzer operator to control an on-line LINC computer. Brecher and Loken (131) show a photograph of a console used to enter sample and test numbers and to send or receive messages to and from a PDP-8L computer. Burns and Bass (154) discuss a remote console designed for AutoAnalyzer-computer-operator communications. The console contains a keyboard through which the AutoAnalyzer operator sends responses and commands to the computer, a Victor eight-column printer to receive numeric results of assays, and a computer-controlled projector-microfilm system which displays preselected texts, such as "How many standards," on a screen.

C. Programming

1. General Considerations

Westlake et al. (899) present a thoughtful discussion of the advantages of discrete automatic analyzers over continuous flow analyzers in relation to computer processing of sample data and of sample identification.

Abdullah (1) describes ALGOL programs applicable to data from peaks when a Sampler module is used or to data from continuous monitoring, i.e., when the sample is aspirated uninterruptedly and no Sampler module is used. The programs incorporate a correction for any scale expansion in use.

Alber et al. (14) discuss software developed for a Digital Equipment Corporation PDP-12A computer with 8 K of core. Data from various AutoAnalyzer channels were logged in and stored on magnetic tape. When all the analytical runs had been completed, the data logging program was replaced by a data processing program which read the taped data, performed the calculations, and produced the final reports.

Young (927) recorded AutoAnalyzer data on punched paper tape for subsequent off-line computer processing. Large amounts of paper tape data were generated, and these tapes were read into an IBM 1130 computer through a high-speed paper tape reader. Young describes several difficulties encountered when using FORTRAN or IBM assembly language routines to read the data tapes; they concerned the speed at which the reader could be operated and the checking of input data for punching or reading errors. To solve these problems, special assembly language programs were developed. The data read from the punched paper tapes were monitored for errors existing on the punched tape itself or errors occurring in the tape reader device. Since the data had been punched on the tape at 7.5-sec intervals from each AutoAnalyzer channel, erroneous punched codes not following the normal rise and fall of the curves could be detected and ignored. Here again, in reading the punched paper tapes, the custom assembly language routines proved superior to FORTRAN, since the latter language halted the reading process whenever an incorrect character was detected. Young gives a good description of how computer programming was used to crosscheck the actual data tapes as well as separately produced punched cards containing sample codes, the number of samples, check samples, and so forth, to prevent processing data incorrectly.

Young (927) also presents several interesting concepts in computerized laboratory quality control procedures. Short-term quality control was obtained by examining CalComp plots of range and geometric moving average charts. These charts allowed immediate rejection of the results from a given run if the check sample fell outside of predetermined limits. For long-term quality control, the individual check sample values were stored on a disk and, at monthly intervals, the average and variance of these data were computed and compared against data sets from the previous month. Data acquisition from the AutoAnalyzer channels was accomplished by the use of shaft encoders on the Recorders, and sporadic trouble with nonlinear response was encountered. A portion of the computer program plotted histograms of the results from each Recorder, permitting the system operator to spot variations due to a problem with an individual Recorder channel. The shaft encoders and paper tape punches were checked by feeding a ramp signal to the Recorder and then statistically evaluating the resulting punched paper tapes for linearity.

2. Peak Picking

Mechanical and electrical apparatuses designed to detect peaks have been mentioned above in Secs. II, B, 1 and 2. In this section, we assume that no previous peak detection has been performed, and that the input data must be examined by the computer program to pick out the peak values. Several techniques are available to the programmer to perform this task. One obvious approach is to use the change in sign of the slope that occurs as the signal passes through the peak value. Successive data values are compared to the previous data values. As the peak maximum is approached, the differences remain positive. The differences pass through zero as the peak goes by and then become negative as the valley begins to be traced out. As long as the data behave as expected, such a peak picking scheme may be carried out by a very short and efficient program segment. Unfortunately, noise spikes from voltage surges, particles passing through the detector flowcell, or other "sinister forces" can cause such a simplistic peak detection routine to fail with alarming regularity. Thus, the programmer is forced to add additional safeguards, such as counting peaks and comparing the count to the number of expected peaks or repeatedly computing the slope of the changing signal, comparing it to the expected ranges or limits to sift out the spikes on the basis of their abnormally high slope values.

A more sophisticated peak picking scheme, which seems to be gaining more and more acceptance, is the use of "time windows" (199, 253, 345, 349, 798, 897). The idea is to examine the data for the presence of peaks only during the short time intervals when a peak is supposed to occur. For example, if the sampling rate is 60 per hour and if the first peak appears 1 min after timing is commenced, the programmer might arrange to examine data from 0 min 50 sec through 1 min 10 sec, 1 min 50 sec through 2 min 10 sec, etc. Incoming data in all other time intervals may be ignored, and noise spikes occurring during the periods in which the time window is closed cause no harm. It may be seen that two values must be measured or known: the sampling rate and the time at which the first analytical peak is expected.

The initial phasing or synchronization of the computer time window to the AutoAnalyzer(s) may be accomplished by a daily determination of the transit times from the beginning of sampling to the appearance of the peak (199). Gray and Owen (345) obtained the peak delay time at the beginning of each day, or whenever a manifold was changed, and noted that it is important not to change the Proportioning Pump end block tube tension after measuring this parameter. Evenson et al. (253) used the signal from the first peak to phase the computer time window; the timing was commenced when the signal dropped away from the first peak.

Once the peak delay time has been measured and the computer timing system has been phased to the AutoAnalyzer, some procedure must be provided to maintain the desired synchronization. AutoAnalyzers exhibit small but unhelpful time wobbles, and several solutions to these timing drifts have been reported. Gray and Owen (345) mounted a microswitch inside the Sampler II which notified the computer each time the Sampler probe moved from the wash box to a cup and vice versa. This switch also started the computer timing system. Ten seconds before each peak was expected, the peak picking routine was started and was allowed to continue for 20 sec. These authors also commented on the timing inconsistency of the Sampler II cam-microswitch mechanism and described the optional alternative use of an external timer. Griffiths and Carter (349) programmed a computer to print an error message if the highest value in a data block did not fall inside the correct time window. Werner (897) and Stewart et al. (798) adjusted the time window phasing, via the computer program, to keep the windows centered on the AutoAnalyzer peaks.

Further comments or discussions on peak picking may be found elsewhere (1, 14, 112, 336, 499, 535).

3. Baseline Corrections and Drift Control

Kirschman et al. (483) discuss a program written in the conversational mode which requires the AutoAnalyzer operator to enter as a variable parameter a noise threshold, that is, a value representing how noisy the baseline is expected to be.

Several programmers make the assumption that baseline drift is a linear function of time (142, 217). Whether or not this assumption is valid, no other model seems to have been proposed in the literature. In order to observe the baseline between each peak, and thus to keep an accurate account of it, the sampling rate must obviously be reduced to the point where the recorder actually returns to baseline on each wash portion of the sampling cycle. This solution is economically unfeasible to a laboratory with large numbers of assays to complete each day. Alsos et al. (20) compared results obtained when the baseline corrections were interpolated between frequent observations (that is, the baseline was measured before and after only a few intervening peaks) and results obtained when the baseline was interpolated between less frequent observations with many intervening peaks. They concluded that frequent observation of the actual baseline is preferable to observations spaced over longer periods of time.

When one is writing a program to deal with system drift, a clear distinction must be made between baseline drift and drift in the sensitivity of the method. Bennet et al. (72) demonstrate through mathematical proofs that, when drift is due solely to a change in baseline, the drift correction is independent of the assay value to be corrected, but if drift is due to a change in the sensitivity of the method, the correction must be scaled according to the assay value to be corrected. These authors used several cups of drift control standard interspersed throughout the run, a practice also recommended by others (897, 927). Stewart et al. (798) corrected for drift by calculating the samples based on the values of bracketing standard cups. The difference in these values was interpolated on the basis of each sample's position between the standards. Gray and Owen (345) also used a linear drift correction based on interpolation between pairs of drift standards. Corrections were applied to the sample values and to the calibration standards lying between the pairs of drift standards.

Programs have also been written to warn of excessive drift. Blaivas and Mencz (113) placed a control sample at the first and last cup position of each sample tray. The computer program corrected each sample value for baseline drift. In addition, it checked the value of the baseline drift against that of the control samples and rejected the run if reasonable agreement was not found. Kirschman et al. (483) described a very flexible program with which the sequence of samples and standards on a sample tray could be inputted before or during the analytical run. Among other things, this permitted selecting which control sample would be used to check for drift. The program flagged the results of the control samples if the drift exceeded the tolerance inputted by the operator.

Other papers contain additional discussions or comments on baseline correction and drift control (21, 199, 252, 336, 349, 499, 535).

4. Calibration and Treatment of Standard Data

Blaivas and Mencz (114) describe a computerized AutoAnalyzer system which included the restandardization of the AutoAnalyzer through calculation of the calibration curve for each new tray of samples. Evenson et al. (252) recommend frequent restandardization of computerized Auto-Analyzers rather than recalculation of results to compensate for drift.

Young (927) calibrated an AutoAnalyzer-computer system by using three standards of different strengths. The first nine cups were filled with triplicate portions of each standard, three of the first, three of the second, and three of the third. The second and third peaks of each triplicate were averaged and used to calculate the standard curve. At the end of the run, the same sequence was employed but in the reversed order of standard strengths.

Several authors (142, 535, 798) have described programs that detect and warn of samples falling outside the range of the calibration curve.

If the drift of the AutoAnalyzer system is low and if the sensitivity is reasonably constant, sample values may be calculated on the basis of the bracketing standards (199). Davidson et al. (217) averaged standards placed at the beginning and end of a run and used the average standard value throughout. Kirschman et al. (483) arranged a computer program to generate straight lines between adjacent standards and used the values of this line to calculate the corresponding samples. Their program permitted inputting the order of cups before or during a run, which allowed correction for errors in placing the cups on the Sampler tray and omitting erroneous peaks from the calculations. Gray and Owen (345) utilized the nearest four standards to yield a four-point interpolation line rather than a two-point interpolation between the nearest two standards. The four-point interpolation process reduced errors arising from random variation in the standard peak heights.

A further refinement in the treatment of standard data consists of regression analysis, usually done by the well-known method of least squares (21, 535, 798). Let us first consider the case for AutoAnalyzer systems yielding fairly linear responses to changes of concentration. The standard plot can then be represented by the usual equation for a line:

$$y = mx + b \tag{1}$$

where y = value of the peak, in absorbance or some other function linearly related to concentration

m = slope of the standard plot line

x = concentration of the standard producing the peak

b = residual (y intercept)

When the equation is used in this form, one may process a series of standards of various concentrations through the AutoAnalyzer and use a least-squares linear regression treatment to obtain the values of m and b. It is best to let x be equal to the more exactly known variable (concentration) and to let y be equal to the less exactly known variable (peak response). Once the values of m and b are computed, the equation is rearranged to solve each unknown peak for x, the concentration of the sample, given y, the value of the corresponding peak. Stewart et al. (798) used a program which performed regression line analysis and calculated the samples versus this line. For systems not linear enough to conform to such an analysis, the program had the option of using a point-to-point curve construction from the standard data.

There is another way to apply linear regression analysis to Auto-Analyzer data. Let us now assume we have satisfied ourselves that the analytical train is yielding data linear with concentration. We then may provide a series of standard cups, regularly interspersed among the sample cups and each containing the same concentration of standard solution. Equation (1) is then rewritten:

$$y = mt + b$$

where t = time (or a numerical value representing the position of the standard cup(s) on the sample tray), and y, m, and b are as previously defined. If there is no drift and no change in the AutoAnalyzer sensitivity, b should represent a statistical estimate of the sensitivity of the method (peak value) at the concentration of standard chosen. Since the same standard solution appears in each standard cup, the linear regression line should be nearly parallel to the time axis, and m, the slope, should nearly approach zero. It can be seen that here the standard line is really measuring the sensitivity drift, and perhaps baseline drift, of the system. A combination of the two approaches (calibration by linear regression analysis of the system's response to various standard strengths, and drift measurement by linear regression analysis of the system's response to an unvarying standard strength over a period of time) provides a fairly reasonable method for treatment of standard data.

If the standard plot is slightly nonlinear, an equation given by Ko and Royer (494) may be useful:

$$1/y = (A/x) + B \tag{2}$$

where x and y are as previously defined. These authors found that Eq. (2) gave a better curve fit than Eq. (1) when used in computer processing of their standard data.

When an AutoAnalyzer calibration curve is definitely nonlinear, more complex equations may be required to obtain accurate curve fits. Keay (458) discusses the advantages and disadvantages of quadratic equations in the use of computers to fit standard curves to digitized data. It was found preferable to convert transmittance data to absorbance (or a related function) before attempting to do the curve fit. Keay gives graphical examples of such curve fits, as computed versus actual data, before and after the data had been transformed from transmittance to absorbance.

The next step upward in complexity is the use of third-order polynomials for the least-squares regression analysis. Bennet et al. (72) used an equation of the form

$$A = a + bP + cP^2 + dP^3$$

where A = assay value, P = recorder peak value, and a, b, c, and d are constants evaluated by the computer program. Bennet et al. discussed the application of this concept to several different clinical analyses which produce variously shaped standard curves. When third-order polynomials were used to fit standard curves, four drift standards of different concentration were required; that is, the calibration curve had to be redetermined at intervals.

Before leaving this topic, we mention an article by Henderson and Gajjar (387) concerning the calibration of nonlinear instruments in general. Their paper stresses the use of digital computers and the least-squares method to obtain curve fits.   The values of unknowns (i.e., samples) are solved either by the computer or by reference to computer-produced calibration tables. Other papers discussing or mentioning the treatment of standard data in computerized systems have been published (1, 113, 349, 897).

5. Data Smoothing and Effects of Noise

In order to function properly, a computer program written to handle data from AutoAnalyzer systems must include some means of dealing with noise. We assume that proper cable shielding or other methods of electrical noise interference rejection are in effect, and restrict the present discussion to noise inherent in the signal from the AutoAnalyzer itself.

The frequency of an AutoAnalyzer signal is extremely low, usually in the range of 1.39×10^{-3} to 1.67×10^{-2} Hz (5 to 60 samples per hour). Superimposed on this slowly changing signal are various unwanted waveforms, some more or less regular in nature, others random in occurrence. Sources of repetitive noise waveforms include the pulsing due to the Proportioning Pump, changes in flow rate arising from the Heating Bath thermostatic cycling action, transients caused by the switches in the Sampler module, and perhaps small amounts of 60-Hz hum from the detector amplifier output. Random noise may be caused by voltage transients from the electrical power source and particles or bubbles passing through the detector flowcell.

Perhaps the simplest approach to data smoothing lies in averaging a certain number of successive incoming data points and then using the averaged points in the ensuing computations. For example, if the Auto-Analyzer signal were sampled every 0.1 sec, successive groups containing 10 data values could be averaged to yield an array of "1-sec" data points. This approach is attractive if core space is limited, because the incoming data may be averaged and then dumped, only the averages being stored. On the other hand, such a noise-smoothing method tends to distort the data by "averaging down" on the upswing and "averaging up" on the downswing. The rate of raw data acquisition must therefore be fast enough that the averaging error is small compared to the instrument error. Furthermore, an averaging process has only a limited ability to deal with noise spikes. A spike appearing in the data block containing the desired peak value can cause an appreciable error, and the magnitude of this error is a function of the width of the spike as well as its absolute value. Returning to our example above, a narrow spike less than 0.1 sec in duration could be missed entirely if it occurred between sampling times. If the spike were read, at the worst its influence would be equal to or less than one-tenth of its maximum value. A broad spike, however, such as one occasioned by a bubble passing through a flowcell, could last several tenths of a second,

TABLE 1

Successive Absorbance Data

	Narrow spike	Broad spike
	0.498	0.498
	0.501	0.501
	0.502	0.502
	0.499	0.768
	0.500	1.000
	1.000	0.864
	0.502	0.602
	0.498	0.498
	0.497	0.497
	0.502	0.502
Average	0.5499	0.6232
Error	0.0499 (9.98%)	0.1232 (24.6%)

and its influence would be much more damaging. This may be illustrated by considering the data strings given in Table 1. Assume the Auto-Analyzer has reached steady state, the true value of the peak is 0.5 absorbance, and the value of the spike (above the true signal) is 0.5 absorbance. Ten successive readings could be averaged as shown in Table 1.

Similar calculations demonstrate that, before a simple averaging technique could reduce such an error to tolerable levels, very large numbers of successive data values would have to be averaged. To do this, and still avoid the averaging distortion mentioned earlier, very high signal sampling rates must be employed. If many AutoAnalyzer channels are to be monitored simultaneously, the required data acquisition rate may exceed the capabilities of the computer multiplexer or may require more core than can be conveniently allocated to the task. Therein lie the weaknesses of simple averaging algorithms as applied to AutoAnalyzer data smoothing.

A much more powerful technique involves the progressive examination of consecutive data subsets, applying least-squares or other curve-fitting processes to adjust the data into a smooth curve. Consider one complete peak from an AutoAnalyzer running at 60 samples per hour. If the signal had been sampled every 0.5 sec,   120 data points would have been acquired. (If we had used a time window the number of data would have been reduced considerably.) We could then apply a least-squares regression to data points 1 through 10. This would yield an equation which statistically best fits these 10 points. We could then calculate the "theoretical" values of the points from the equation and place these smoothed data back into the array, wiping out the 10 original data points. Next, we would examine data points 2 through 11 (the first nine having already been smoothed once) and continue applying the algorithm (3 through 12, . . . , 111 through 120).   Such a successive, repetitive smoothing process deals much more effectively with spikes than the simple averaging technique described earlier. There are several arguments against these procedures and, as always, one must decide whether the tradeoffs are worth the enhanced quality of data. Obviously a greater programming effort is required, and longer programs further reduce the available core space for storage of data. If one wishes to run on-line, the computer will have to do the data smoothing as the data come in, dumping the data after arriving at the final desired value so as to make core space available for fresh incoming data. Let us assume we are reading data at 0.1-sec intervals from each of 16 AutoAnalyzer channels. Let us further assume that about 150 separate addition, subtraction, multiplication, or division operations are required for one pass through a least-squares regression analysis of 10 data points. To keep ahead of 160 data points per second, the computer will have to perform 24,000 operations per second for data smoothing only. Although this rate is within the capability of many central processing units, there may not be much time left to load and run other desirable and necessary operations, such as data acquisition and report writing programs. Thus, one may be forced into a combination of on-line data acquisition and later off-line data processing.

Gray and Owen (345) acquired data at 40-msec intervals from each of five AutoAnalyzer channels. Consecutive readings were averaged in sets of eight, and a second-order exponential smoothing process was then applied to the averages. Alber et al. (14) logged data onto magnetic tape once every 3 sec from each AutoAnalyzer channel. After the analyses were completed, the data were read from the tape and were smoothed through an eight-point, least-squares algorithm. An oscilloscope display allowed the operator to observe the logged data before and after smoothing.

Evenson et al. (253) studied the relationship between sample to wash ratio, noise, and precision while acquiring data from AutoAnalyzers with a computer. An AutoAnalyzer system exhibiting a noisy steady state was selected, and data were acquired while the system was operated at a constant rate, 60 samples per hour, with 1:2, 2:1, and 11:1 sample to wash ratios. The observed precision was much better at the highest sample to wash ratio, even though this ratio made the noise the most visible on the recording. These authors feel that precision will be better if high values of sample to wash ratios are used, even in relatively noise-free AutoAnalyzer systems.

Articles by Werner (897) and Krichevsky et al. (499) contain further information on methods for dealing with noise in computerized Auto-Analyzer systems.

6. Corrections for Carryover (Sample Interaction)

The theoretical aspects of carryover are thoroughly discussed in Chapter 7. In this section, we shall confine ourselves to articles pertaining to the use of computer programs in correcting for carryover.

Young's method (927) of calibrating computerized AutoAnalyzer systems has been previously mentioned. To repeat, the first nine cups were filled with triplicate portions of three different standard solutions: three of the first, three of the second, and three of the third. The first peak of each triplicate was compared to the second and third peaks to estimate the carryover, and a correction was then applied to the sample values via the computer program. At the end of the run, the same sequence of standards was employed, but in the reversed order of standard strengths.

A program reported by Kirschman et al. (483) required the Auto-Analyzer operator to enter, along with other parameters, the percent correction for carryover.

Bennet et al. (72) defined percent carryover by the equation $C = 100(A_2 - A_1)/A_1$, where C is percent carryover, and A_1 and A_2 are assay values corresponding to two sequential peaks, i.e., the peak causing the carryover and the peak directly following it. To study carryover, these authors used their aforementioned computer-Sampler II interface to control the Sampler probe action accurately. AutoAnalyzer systems were operated at 90 samples per hour to emphasize carryover, and the effect of the assay value of a sample causing carryover upon the assay value of the next sample was studied in ten different clinical assays. In general, the percent carryover was found to be independent of the assay value of the peak causing the carryover, and this justified the use of the proposed proportional correction. Furthermore, the percent carryover was found to be the same at 1:1 or 3:1 sample to wash ratios at the same sampling rate. However, as the sampling rate was increased, percent carryover increased.

In the same year, Evenson et al. (253) also published a study of carryover in computerized AutoAnalyzer systems. To emphasize carryover, these authors lowered the flow rate through the flowcell. As the sampling rate was increased, the percent carryover also increased. The measured value of the carryover agreed well with the value calculated theoretically by the method of Thiers et al. (832, 833). Next, Evenson et al. studied how a change in the sample to wash ratio affected percent carryover at a constant sampling rate of 60 per hour. Data were taken at sample to wash ratios of 11:1, 5:1, 2:1, 1:1, 1:2, and 1:5. Within experimental error, the percent carryover was not affected when tested with a sequence of standard solutions containing 0, 50, 100, and 50 mg glucose per 100 ml. In these tests, the peak value as such was not used; instead, values were obtained from computer-timed readings corresponding to the Sampler rate after observing the time of appearance of the first peak. As we have already noted, these authors found that precision (in noisy or noise-free systems) was best at the highest sample to wash ratio. They point out that the major objection to using a high sample to wash ratio seems to be an intuitive feeling on the part of the AutoAnalyzer operator that the attendant lack of "wash" will somehow increase carryover. Their experiments demonstrate that this feeling is unjustified. By selecting conditions that give good wash at the expense of not approaching steady state, one falls prey to timing errors and "becomes blind to problems of noise."

Thiers et al. (834) put these practical and theoretical observations to work in a computer program designed to correct for carryover in AutoAnalyzer systems. A sequence of seven cups was assayed to acquire the necessary data: five standards in decreasing order of concentration, highest to lowest, then a cup of blank, and finally a cup of the middle standard concentration. The computer used these data to form a corrected standard curve through a series of approximations and then calculated each sample value, corrected for carryover. The percent carryover was also calculated. The algorithm used in the computation is outlined in their paper. To validate this method of carryover correction, data from numerous experiments were collected and presented graphically, uncorrected and computer-corrected results being used for comparison. Accuracy and precision were dramatically improved by the use of the correction algorithm. The percent carryover could be increased to 15% or more without serious deterioration of the data when the corrections were applied. As an additional test, a manifold was purposely set up to yield about 8% carryover by adjusting the flow rate through the flowcell. To determine whether the order of standards affected the corrected calibration curve, standards were passed through the AutoAnalyzer in ascending, descending, and random order. The uncorrected calibration curves showed distinct differences, but the corrected calibration curves remained constant.

For further comments on carryover in computerized continuous flow systems, the reader may consult other papers (113, 114, 142, 199, 899).

7. Diagnosis and Flagging of System Malfunctions

We have touched briefly on this topic in the previous sections, for example, the flagging of samples whose values fall outside the range of the calibration curve. Blaivas and Mencz (112, 113) placed a control sample in the first and last cup of each sample tray, and the computer checked for the appearance of the control values at the expected intervals to ensure that the sample tray had been inserted correctly on the Sampler module. The values of the control samples themselves were also checked by the computer against preset tolerances.

Criteria used in the detection of system malfunctions usually fall into either or both of two categories: peak timing and peak shape. In a program described by Kirschman et al. (483), the computer started the Sampler II rather than the AutoAnalyzer operator. The computer was thus allowed to measure the delay time from sampling to peak, and this datum along with peak and valley detection permitted the computer to check on the validity of each peak. Levy and Kanon (535) used a program which examined each peak for acceptable shape and timing. Evenson et al. (252) monitored the proper operation of AutoAnalyzers by programming measurements of standard drift, variation of times between peaks, and the uniformity or scatter of the readings taken at the peak. The latter measurement would detect an irregular peak or perhaps a bubble in the flowcell. These authors recommend that the computer be programmed to signal the AutoAnalyzer operator if a malfunction is detected so that immediate inspection of the Recorder chart is possible.

Blaivas and Mencz (112-114) developed computer programs to detect AutoAnalyzer malfunctions based on the assumption that the peaks should be parabolic in shape. To be accepted, each peak had to be smooth, parabolic, and of a time interval corresponding to the sampling rate. Each peak reached steady state in their AutoAnalyzer system, and the peak plateau time could thus be examined to reject peaks arising from noise or from insufficient volume of sample in a cup. A peak following a sample of very high value (i.e., one exceeding a preset limit) was also rejected. Blaivas and Mencz (114) carried the "parabolic model" to its logical conclusion by stating that the area under an AutoAnalyzer peak is proportional to its concentration, and thus the peak height is proportional to both area and concentration. The computer was therefore programmed to reject a peak if its shape was not parabolic or if its height was not proportional to its area. *

---

*AutoAnalyzer peaks may at times appear to be parabolic in shape, but this mathematical model is not entirely satisfactory (see Chapter 7).

Further refinements in the diagnosis of AutoAnalyzer malfunctions may be applied when using programs which acquire data only during time windows corresponding to the expected arrivals of peaks. Stewart et al. (798) adjusted the time window via the program to track the AutoAnalyzer timing drift, that is, to keep the data sampling intervals centered on the peaks. However, if the AutoAnalyzer timing drift exceeded the capability of the program to track and correct for it, the program issued a warning. The program also monitored standard drift (which could indicate either excessive AutoAnalyzer drift or an incorrectly prepared standard), the order of standards (the correct placement sequence on the sample tray), and the conformance of quality-control samples to prescribed limits. Griffiths and Carter (349) used their program to check the data block containing the peak for continuity of rise and fall in the values about the maximum and to print an error message if significant deviations from smoothness were found. Evenson et al. (253) proposed a very interesting concept for computerized diagnosis of AutoAnalyzer malfunctions. Seven data readings were taken during a time window, based on the Sampler rate, starting just before the time of the expected steady-state peak and ending just before the dropoff from the peak. Normally these seven data would rise to a steady value and hold there. Noisy steady states yielded up-down-up-down changes. Excessive carryover leading to a shoulder peak on the tail of the preceding peak yielded a down-down-down pattern. A noise spike yielded a rapid-up-then-down pattern. Sampler timing variations, improper synchronization, Sampler cam jump-ahead, and clogged manifold lines also gave distinctive patterns in the seven values obtained in a time window; thus, rather specific diagnoses of system malfunctions were possible.

Werner (897) and Poletti et al. (665) also give discussions on the elimination of incorrect results from computerized AutoAnalyzer systems.

D. Special Applications

We conclude this chapter by mentioning four interesting and rather specialized uses of computers with continuous flow analytical systems.

Roodyn and Maroudas (693) used a gradient formed by the use of a vessel and mixer (see Chapter 4, Sec. II, L), along with an AutoAnalyzer, to study enzymes. These authors describe the required calculations, which were found to be tedious, and a FORTRAN program which was used instead. The input to the program was from punched cards. Pitot et al. (662) studied enzyme reaction rates with an AutoAnalyzer and collected the data via a digital voltmeter, an output adapter, and an IBM card punch. The data were then processed by a computer; the program is described in their paper.

Mor et al. (604) measured glucose, ethanol, amino nitrogen, and ammonia in yeast cultures.   These authors outlined a FORTRAN IV program which used the data acquired from the various AutoAnalyzer channels to calculate several culture parameters, including specific oxygen uptake, specific carbon dioxide release, respiration quotient, carbon balance, substrate turnover, and dilution rate.

Frazer et al. (288) employed an AutoAnalyzer manifold as a part of the determinative step in an analysis of ammonia in water.   The ammonia was distilled manually, and the distillate was assayed colorimetrically by the AutoAnalyzer manifold and a double-beam spectrophotometer interfaced to a PDP-8/S computer.   This system did not measure a peak value. Rather, through a voltage-to-frequency converter and a counter controlled by the computer, it integrated the signal produced by the spectrophotometer. The entire collected distillate was pumped into the manifold over a period of about 285 sec; thus, all the ammonia in the distillate was measured in the integrated signal.   This technique gave a marked increase in sensitivity compared to the more common sampling techniques, e.g., the use of a Sampler II.   The precision, expressed as relative standard deviations, ranged from 9.8% at the 0.4-μg level to 1.7% at the 8-μg level.

Chapter 7

THEORETICAL ASPECTS

W. H. C. Walker
McMaster University
Hamilton, Ontario, Canada

I. INTRODUCTION

Since the introduction of continuous flow analyzers in 1956, there has been much experimental activity directed toward method development. With great ingenuity, many manual analytical techniques have been adapted to continuous flow operation. By contrast, research into basic aspects of the functioning of the machine itself has been very limited. Such insight may have been available within the industry since rational equipment redesign has led to a steady improvement of performance. Nonetheless, with any given machine, optimum performance is dependent not just on trial and error but ultimately on the application of certain general principles relating to the continuous flow concept. Credit must go to Thiers and his collaborators for initiating research in this area and, in a series of lucid papers (253, 357, 832, 834, 835, 835a), for contributing to every major advance. Practical and theoretical aspects of this area have recently been reviewed (830a).

In this chapter, two topics are considered: first, the factors that determine peak characteristics and, second, the practical steps that may be taken to improve peak quality and analytical performance.

A perfect continuous flow system would yield an output trace that corresponded exactly with input solute concentration changes, subject only to a time delay appropriate to the chemical reactions involved. In such a system, sampling rate would be limited only by sampler mechanics and recorder response time. Transfer of sampler probe from wash to sample and back to wash would result in a sudden square-wave transition of the recorder pen from baseline to plateau and back to baseline.

In practice, such a square-wave output is never seen. A sudden change of concentration at the sampler probe results in a gradual output transition, sigmoid in outline, from one steady state to another. The fluid stream linking sampler to colorimeter imposes a deformation on the input solute concentration change, and this deformation is the limiting factor in sampling rate; it affects the extent to which peaks approach plateau, the magnitude of interaction between one peak and the next, and the frequency of shoulder peaks in which a high sample obliterates the peak of a following low sample.

There is good experimental evidence (835a, 884) that the deformation has two major components, one related to the segmented stream and the other related to the unsegmented stream in debubbler and flowcell.

II. SEGMENTED STREAM DEFORMATION

If a single pulse of solute molecules enters the segmented stream after passage through a short sample line and sample pump tube, it will start its journey along the segmented stream much as it entered the sample probe, as a compact zone of solute occupying one or at most two fluid segments. When it arrives at the end of the segmented stream, the compact zone will have broadened and, in the absence of adsorptive effects, will have a symmetrical falloff in concentration ahead of and behind the zone of maximum concentration. This effect may readily be observed by using bromocresol blue in the sample cup.

It will be apparent that this symmetrical dispersal applies only to the segmented stream. The recorder output will show a peak whose asymmetry results from the effects of the debubbler and flowcell. If the experiment is repeated with the air line clamped, the efficacy of the air bubbles in restricting the magnitude of the zone broadening will be apparent.

The deformation due to the segmented stream has been measured experimentally in three ways. Fleck et al. (280a) collected each individual segment and measured its absorbance in a spectrophotometer. Thiers et al. (835a), by using a flowcell with a bubble sensor to suppress the bubble artifact at the recorder (357), passed the segmented stream through the flowcell without debubbling. Walker et al. (883, 885) used a flowcell with debubbler and removed the superimposed flowcell artifact by computation.   The data from each group support the same conclusion: Solute molecules entering the segmented stream travel with velocities that are normally distributed about the mean flow rate, * so that the dwell time of molecules in the segmented stream is distributed normally about the mean dwell time t_D with a standard deviation σ_D.   In consequence, a transition from wash to sample, representing a square wave at the input, emerges from the segmented stream with a sigmoid front described by the integral of the normal probability curve (Fig. 1a). Theoretical models provided by these workers readily yield either Poisson or gaussian distributions. It has further been shown that the Poisson model is little affected by solute-dependent variation in segment to segment fluid transfer (70). With the large number of segments employed in practice, the Poisson and gaussian models are indistinguishable and correspond closely with observed data.

III. FLOWCELL DEFORMATION

The deformation attributable to the debubbler and flowcell can be examined in isolation by presenting the debubbler with a square-wave transition from one steady state to another. The recorder response will indicate deviation from a square wave which is due only to the flowcell. Using an AutoAnalyzer Mark I colorimeter, the steady-state transition is readily achieved (884) by halting flow through the pull-through line from the flowcell while changing solute concentration at the input probe. All the segmented stream passes to waste at the debubbler while the fluid in the flowcell remains undisturbed.   If flow through the flowcell is restored after a new steady state has been achieved at the debubbler, a square-wave transition will be presented to the flowcell independent of the length and nature of the segmented stream.

Under normal operating conditions, the deformation due to the flowcell conforms closely to an exponential curve (835a, 884) (Fig. 1b). A similar exponential deformation is introduced by other unsegmented parts of the liquid stream.   Normally the sample line from probe to pump contributes a small exponential component.   A much larger increment is added if sample is diluted in the manifold and resampled after the diluted stream is debubbled.

---

*This is less than the apparent flow rate of the segments since there is always a proportion of the fluid stream that is stationary on the walls of the parts of the tube occupied by bubbles (830a).

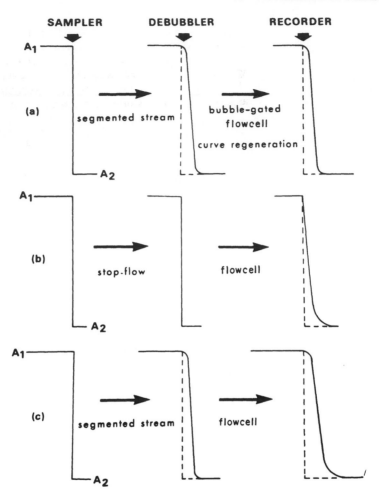

FIG. 1. Square-wave deformation.   (a) Normal probability integral
deformation in segmented stream between sampler and debubbler. Flowcell deformation eliminated.   (b) Exponential deformation in flowcell.
Square wave presented to debubbler by stop-flow technique.   (c) Combined
effect of deformations in segmented and unsegmented streams.

IV. TOTAL DEFORMATION

The dominant deformations are "normal probability integral" in the
segmented stream and exponential in the unsegmented stream.   On a
recorder trace of a sigmoid transition from baseline to plateau (rise
curve) or from plateau to baseline (fall curve), every point on that trace
has been subjected to both deformations acting sequentially (Fig. 1c).

It is not valid, however, to consider the early part of the curve as representing segmented stream deformation and the latter part of the curve as representing flowcell deformation. Such an interpretation is tempting because the latter part of most curves is indeed exponential in nature (832, 884); this is because a new steady state has been attained at the debubbler and flowcell effects alone are then manifest.

The association of normal probability integral deformation with the segmented stream and exponential deformation with the unsegmented stream is a first approximation which agrees closely with observed data from systems operating under normal conditions. If the segmented stream is greatly increased in volume, a small increase in exponential deformation results. If the flowcell pull-through rate is reduced, there are minor increases in normal probability integral deformation (883a). In the absence of adsorptive effects, every continuous flow system tested (nephelometric, fluorometric, flame emission and absorption, and colorimetry, both positive and negative) introduces a deformation which can be resolved into two major components, one normal probability integral and the other exponential.

V. DEFORMATION CONSTANTS

The combined sigmoid and exponential deformation can be defined in terms of two constants: the standard deviation of the dwell time σ_D and the exponential factor b. The extraction of these constants from any rise or fall curve is a simple graphical calculation procedure.

It is necessary first to clarify the effects of the scalar changes in electrical signal that distinguish different colorimeters, older models giving a transmittance signal, newer ones giving an absorbance signal. The exponential factor due to the debubbler and flowcell relates strictly to concentration of solute molecules in the flowcell. Colorimeter output must therefore be converted to an absorbance scale, and the colorimeter must have narrow band pass filters and sufficiently low stray light to ensure conformity with Beer's law over the concentration range concerned. Similar linearity constraints apply to flame emission spectrophotometry and fluorometry. The σ_D value relates to solute molecules throughout the segmented stream whether in their native form or after transformation to a colored end product. Limitation of end-product formation by reagent concentration will invalidate σ_D estimation. There is no requirement for reactions to proceed to completion, providing that all samples proceed to the same extent regardless of their concentration. A combination of these constraints requires that output signals relate linearly to solute concentrations at the colorimeter and that these in turn be proportional to input solute concentrations. Given these provisos, two postulates logically follow and have been experimentally validated (832, 884).

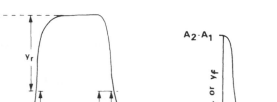

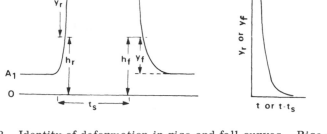

FIG. 2. Identity of deformation in rise and fall curves. Rise and fall curves between steady states A_1 and A_2 are shown together with the same curves expressed in terms of y, the distance the pen has to travel to the next steady state.

First, σ_D and b are independent of initial and final steady-state absorbances. In particular, (a) deformation constants are independent of peak heights, and (b) rise and fall curves between two given absorbance steady states are identical reversals of each other; that is, they are identical when expressed in terms of the distance the pen still has to go to reach its next steady state. This is a useful measure and is denoted by y (Fig. 2). Deviation from this rule implies discrimination between solute and solvent on the basis of solute concentration. For example, the nonlinear isotherm of adsorption will result in trailing of the low concentration end of the fall curve.

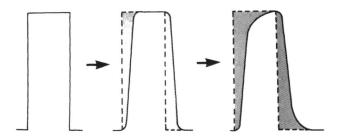

FIG. 3. Parity of areas. The original square wave representing a given mass of colored solute is repeated in dotted outline over the corresponding normal probability integral deformation and over that deformation combined with exponential deformation. In both cases the striped area lost from the square-wave rise curve is added to the fall curve.

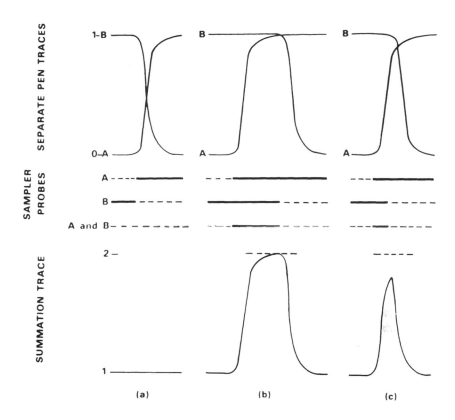

FIG. 4. Additive nature of rise and fall curves. Using a common sample with final absorbance 1.0, two identical analytical channels have their sampler probes inserted in sample over the time intervals shown. The two output traces are superimposed (top) and the arithmetic sum of the traces is shown on the summation trace (bottom). At all times, one probe or the other is in the sample. When only one probe is sampling, the summation trace has absorbance 1.0. When both probes are sampling, the summation reaches 2.0 after deformation effects have disappeared. In (a) the summation of identical rise and fall curves is shown. The curves are not displaced in time and no peak results. In (b), the fall curve is delayed until after the rise curve has reached plateau and a plateau peak results. In (c) the fall curve starts before the rise curve has reached plateau and results in a lower peak whose height is defined by $y_f - y_r$.

Second, the area under a peak is independent of the extent of its deformation and is equal to the corresponding square-wave input (Fig. 3). If a peak does not come to plateau, the peak outline is the arithmetic resultant of the rise curve and its corresponding fall curve displaced by the sampling interval t_s (Fig. 4). Similar additive considerations apply to sequential nonplateau peaks (Fig. 5).

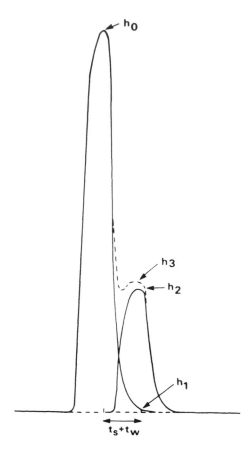

FIG. 5. Additive nature of sequential nonplateau peaks. The low peak, true height h_2, has added to it (broken line) the coincident fall curve of the preceding peak h_0. At a time corresponding to sample time t_s plus wash time t_w, the fall curve of h_0 has a height h_1. Peak h_2 is increased by the same amount to give height h_3.

A. Calculation of b

If a rise or fall curve is plotted semilogarithmically as log y versus t, there is an initial nonlinear relationship and the rest of the curve is linear (832, 884). The early nonlinear part corresponds to the time required for the segmented stream to reach a new steady state at the debubbler.

The linear relationship is defined by the exponential equation

$y = A \exp(-t/b)$

where A is the absorbance interval between initial and final steady state. The exponential constant b defines the magnitude of the exponential deformation. It is defined graphically as the time required to pass from any y value y_t to $0.37y_t$.

B. Calculation of σ_D

In order to measure σ_D from a standard rise or fall curve, the exponential deformation must first be removed. This process is termed curve regeneration (883) and employs the fact that the relationship between any absorbance γ_t at the debubbler and the corresponding absorbance h_t after exponential deformation is given (885) by

$$\gamma_t = b \frac{dh}{dt} + h_t$$

If absorbances are measured on a rise or fall curve at 2-sec intervals, dh/dt is approximated by $0.5(A_n - A_{n+1})$, where A_n and A_{n+1} are consecutive 2-sec absorbance readings. The corresponding h_t is the midpoint of the two readings, $0.5(A_n + A_{n+1})$. If γ_t values, suitably transformed, are plotted on a probability scale against time, a linear relationship results. Since probability plots have limits of 0 and 1, it is necessary to transform γ_t values to γ_t':

$$\gamma_t' = \frac{\gamma_t - A_{min}}{A_{max} - A_{min}}$$

where A_{max} and A_{min} are the two steady states involved.

A working example will best illustrate the process (see Table 1). Consider a fall curve from an absorbance of 1.0 to baseline (absorbance zero). A series of 2-sec absorbance readings (0, 1, 2, ... , n) is obtained using a fast chart speed (transposing driving and driven gear wheels on an AutoAnalyzer Mark II recorder suffices); b is previously defined as 10 sec. Figure 6 shows a plot of γ_t' on a probability scale. The interval between $\gamma_t' = 0.50$ and $\gamma_t' = 0.16$ is indicated. This corresponds to one standard deviation. Hence, $\sigma_D = 4$ sec.

TABLE 1

Working Example of σ_D Calculation

n	0	1	2	3	4	5	6	7	8	9
Time (sec)	0	2	4	6	8	10	12	14	16	18
A_n	1.000	0.985	0.960	0.910	0.840	0.750	0.650	0.545	0.450	0.370
$A_n - A_{n+1}$		0.015	0.025	0.050	0.070	0.090	0.100	0.105	0.095	0.080
$-dh/dt$		0.0075	0.0125	0.025	0.035	0.045	0.050	0.0525	0.0475	0.040
$-b\,dh/dt$		0.075	0.125	0.250	0.350	0.450	0.500	0.525	0.475	0.400
h_t		0.993	0.973	0.935	0.875	0.795	0.700	0.598	0.498	0.410
γ_t		0.918	0.848	0.685	0.525	0.345	0.200	0.073	0.023	0.010
Time (sec)		1	3	5	7	9	11	13	15	17

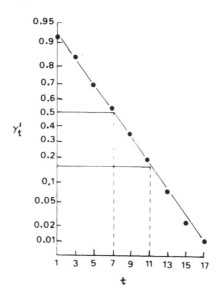

FIG. 6.   Plot of γ_t' versus t.   A cumulative probability scale is used for γ_t'; a linear scale is used for t.   The time interval corresponding to a γ_t' change from 0.5 to 0.16 defines σ_D; σ_D = 4 sec.

C.   Significance of b and σ_D

The magnitudes of b and σ_D closely define the deformation of a square-wave input.   In practice, the magnitude of b plays a dominant role in determining the three important peak characteristics: approach to plateau, shoulder peak frequency, and interaction.

1.   Approach to Plateau

Peak heights of 90, 95, and 99% of plateau have y/A values of 0.10, 0.05, and 0.01, respectively.   From the exponential relationship $y/A = e^{-t/b}$, these peak heights correspond to sampling times t_s of 2.3b, 3.0b, and 4.6b, respectively.   Under most operating conditions, σ_D plays a trivial role and may be ignored (884).

2.   Shoulder Peaks

A shoulder peak results when a low peak follows a high peak within such a short period that the minimum between the two adjacent peaks is eliminated.   If the upward velocity of the rise curve of the low second peak is less than the residual downward velocity of the fall curve of the high first peak, the pen fails to move upward to form a second peak and a shoulder peak results. Pen velocities can be calculated by differentiation of the corresponding exponential equations and it can readily be shown that the occurrence of shoulder peaking is a function of the ratio between first and second peaks and of the wash time t_w between samples. With peak ratios of 2:1, 5:1, and 10:1, shoulder peaks result if the wash time is less than 0.7b, 1.6b, and 2.3b, respectively (884). Again, σ_D deformation plays a negligible role.

3. Interaction

Since the tail of the fall curve of a preceding peak is added to any following peak, interaction is a fixed proportion of the preceding peak height (835) and is a function of the time interval between peaks, $t_s + t_w$ (253, 884). From the exponential equation, it follows that interaction equal to 7, 1, or 0.5% of preceding peak height will occur when the interval between samples is 2.65b, 4.6b, or 5.3b, respectively (884). A sample interval of 2.65b will not only produce interaction equal to 7% of a first peak on the second peak, but will also produce interaction of 0.5% on the third peak. Such extended interaction is hopefully seen only in high-speed sampling systems with computer correction for interaction (253a, 834). At such sampling rates a further effect, termed inductive interaction, may occur (253a). Here, the peak preceding a high peak may show interaction due to the high peak that has still to emerge from the system. This is a σ_D effect and results from the fact that, while some solute molecules move with less than the mean velocity of the segmented stream, an equal number move faster than the mean velocity and can catch up with the sample ahead.

4. Interdependence of Sample and Wash Time

Whereas peak height is determined by sample time and shoulder peaking is determined by wash time, interaction is a function of both these times. It is fortunate that these effects are not in conflict: Increasing sampling time brings peaks nearer to plateau and reduces interaction; increasing wash time reduces shoulder peaks to an acceptable frequency and further reduces interaction. Since a peak height 5% away from plateau is acceptable but interaction realistically must be less than 1%, the added effect of wash time on reducing interaction conforms to practical requirements.

With sampling time set at 3.0b to give peaks 95% of plateau, a wash time of 1.5b eliminates shoulder peaks on all sequential peak ratios less than 4.5:1 and interaction is 1%. With the same sample time, a wash time of 3.0b has a shoulder peak ratio of 20:1 and interaction is 0.3%. If peak heights can be read at precisely timed intervals (as with the SMA 12/60 or with on-line data acquisition), shoulder peaks no longer present a problem and improved peak characteristics result from extending the sampling time to 5/6 or more of the sampling interval (253).

There can be no justification for using sampling times less than 3.0b. The resultant spike peaks mask short sampling due to inadequate plasma volume or intermittent sample line blockage. Despite the fact that peaks often look "better," imprecision due to inherent noise is amplified (253). Moreover, unless computational corrections are made, interaction will always be increased to unacceptable levels unless wash time is greatly lengthened. The latter situation is present in any system in which sample cups are alternated with water cups. Unless sample volume is at a premium, such systems would function better with a regular 2:1 sample-wash ratio and a halving of the sampling rate.

5. Further Comments on Interaction

Two definitions of interaction are currently employed. The first (835) states that interaction is a constant proportion (k) of the preceding peak height. The second (36a) states that interaction is a constant proportion (k') of the difference in the height of the preceding peak and that of the peak subjected to the interaction. Although in apparent conflict, both statements can be correct with the constants k and k' equal (835, 884). Which definition is appropriate depends only on the way peak height is defined.

The matter is best illustrated by the example of three identical samples run in sequence after a period of water wash (Fig. 7). The three peaks have heights h_1, h_2, and h_3 and, in most systems, h_2 and h_3 are identical, since appreciable interaction usually does not extend beyond the adjacent peak.

If h_1 is defined as the true peak height, then h_2 and h_3 carry an increment due to interaction which is a fixed proportion of h_1 and corresponds to $h_1 \exp(-p/b)$, where p is the time interval between peaks (884).

On the other hand, if h_2 is defined as the true peak height, h_3 carries zero interaction, h_1 carries a negative interaction equal to $-h_1 \exp(-p/b)$, that is, $(0 - h_1) \exp(-p/b)$, and in general any peak h_n carries interaction equal to $(h_{n-1} - h_n) \exp(-p/b)$.

The choice of which peak height is "true" is arbitrary. Neither conforms to the usual operating practice of analyzing series of standards of ascending or descending concentrations. The first situation can readily be achieved by running a water sample between each standard; the second, by running each standard in duplicate. In any system that generates plateau peaks, interaction becomes zero (306) and the definitions are redundant.

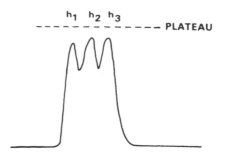

FIG. 7. Interaction. Three identical peaks have apparent peak heights h_1, h_2, and h_3. Peak heights h_2 and h_3 are identical; h_1 is lower because it is preceded by water. If h_1 is used as a datum, h_2 and h_3 both carry interaction which is a constant proportion of h_1. If h_2 is used as a datum, h_1 carries negative interaction of the same magnitude; h_3 carries no interaction. In general, when the first datum is used, interaction is proportional to preceding peak height; when the second datum is used, interaction is a similar proportion of the difference between the height of the peak in question and that of the preceding peak.

VI.   OPERATIONAL FACTORS INFLUENCING DEFORMATION CONSTANTS

In most current analyzer systems, sampling rate is directly dependent on the magnitude of b. It is therefore necessary to define the ways in which b can be minimized. There will be increasing use of systems in which exponential deformation is effectively eliminated. In these systems, the magnitude of σ_D becomes the determinant of sampling rates. The need to process increasing numbers of samples is not the only justification for reducing square-wave deformation. Along with higher sampling rates goes reduced sample volume requirement and reagent consumption, the latter being especially relevant in high-cost enzyme assays and expensive specific antiserum usage on the specific protein analyzer. If none of these considerations applies, reduced deformation provides extended periods of plateau output with commensurate reduction of noise-induced imprecision.

A.   Minimization of b

Three choices are available: (a) improving wet flow conditions on existing equipment, (b) electronic removal of exponential deformation, and (c) equipment redesign.

1. Wet Flow Conditions Affecting b

The volume of the unsegmented stream must be reduced to a minimum. The unsegmented sample line should not be unduly long and should have a bore no larger than that of the sample pump tube.

Dilution of sample on the manifold with debubbling of the segmented stream and resampling of the unsegmented residue is no longer acceptable. The Technicon A10 fitting, which consists of a glass tube with a fine metal tubular insert penetrating to the center of the tube and presenting a bevelled edge to the oncoming segmented stream, permits direct re-sampling of the diluted stream without debubbling. Such dilution on the manifold increases σ_D but does not affect b.

Values of b are reduced by increasing the pull-through rate through the flowcell (306, 357, 884). There is diminishing return for increasing rates, and 5.0 ml/min and 1.0 ml/min represent reasonable maxima for AutoAnalyzer colorimeters Models I and II, respectively. The fluid volume in the waste line from the debubbler represents noncontributory information and should be reduced to the minimum consistent with efficient debubbling. In practice, the segmented stream liquid flow rate should not exceed 1.4r, where r is the pull-through rate.

Discrepant donor and recipient dialyzer flow rates are not acceptable. High-viscosity reagents increase deformation and should be avoided if possible.

2. Electronic Devices

A square wave between two steady states differing by an absorbance A undergoes an exponential deformation at the debubbler,

$$y = A \exp(-t/b)$$

where y is the distance away from the new steady state. The slope of this curve, dy/dt, is given by

$$dy/dt = (-1/b)A \exp(-t/b) = -y/b$$

Hence, $y = -b\, dy/dt$. It follows that the distance the pen has to go to reach its next steady state is predictable before that steady state is reached (832). Simple electronic circuits can measure slope, multiply by a constant, and add the result to existing pen height (886). This process, termed curve regeneration, is generally applicable to exponential deformation regardless of the shape of the input curve. Curve regeneration units are

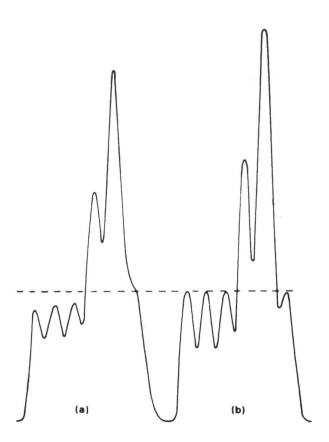

FIG. 8. Curve regeneration, BUN method. Samples: 0.25, 0.25, 0.25, 0.5, 0.75, 0.25 g/liter; σ_D = 2.5 sec, b = 7.5 sec, t_s = 20 sec, t_w = 4 sec (150 samples per hour). Broken line indicates corresponding plateaus. (a) Without regeneration; (b) with regeneration.

commercially available (Fisons Scientific Ltd., Loughborough, Leicestershire, United Kingdom) and can be directly interposed between colorimeter and recorder without any equipment modification. With the use of such a device, sampling rates can usually be doubled with no loss of peak quality (166a) (Fig. 8). An input signal of high quality with no excessive noise is required. The device makes good peaks better; it does not make bad peaks good. Curve regeneration can be viewed as putting into quantitative terms the observation familiar to all AutoAnalyzer users that a high peak is obvious long before the apex is reached because the pen is moving faster. By making use of the data contained in pen velocity, more of the available information is being captured with consequent improvement of precision or reduction of sample volume.

3. Equipment Redesign

Substantial reductions in exponential deformation have been achieved by the flowcell redesign of the AutoAnalyzer II colorimeter. Further improvements are evident in the flowcell of the SMAC system, and remaining exponential deformation in this system is largely removed by passage of an undebubbled stream through the flowcell with digital computer processing of the output signal to eliminate the optical artifacts due to bubbles. This is the logical development of the bubble-gated flowcell (357) in which conductivity changes within the flowcell were used to suppress bubble optical effects. An analog device which identifies bubble artifact and removes its effects from the output signal has been described (623a). Use of such a device is logically sound and results in greatly improved quality of output on AutoAnalyzer II systems with reduction of sample and reagent volumes.

B. Minimization of σ_D

The possibilities are the same as those for b. The magnitude of σ_D becomes dominant only in systems that use curve regeneration or bubble gating. Nonetheless, it is reasonable to minimize deformation on any system.

1. Wet Flow Conditions Affecting σ_D

It has been shown (883a) that σ_D is related to (a) the bubble interval t_B, that is, the time in seconds for the passage of one air plus liquid segment; (b) the air-liquid ratio (a/l), which is dimensionless with both a and l defined in milliliters per second; and (c) the volume V of the segmented stream (in milliliters) (835a). The relation is given by the equation

$$\sigma_D^2 = kt_B(a/l)V$$

Since $V = t_D(1 + a)$ and $t_B = V_B/a$, where t_D = dwell time (seconds) and V_B = mean volume of a single bubble (milliliters),

$$\sigma_D^2 = kV_Bt_D[1.0 + (a/l)]$$

The magnitude of V_B should be reduced to the limit of bubble stability. The value of t_D is determined by reaction time, and faster reactions are to be preferred wherever a choice exists. The term a/l should not exceed 0.5. The value of t_B must be less than $0.5\sigma_D$ if mixing that occurs within segments is not to become a dominant effect (883a). Reduction of a/l does imply ultimately an increase in fluid flow rate, and this requires compromise since increasing flow rate has implications for bubble stability, backpressure, reagent consumption, sample volume, and sensitivity. The nature of the tubing containing the segmented stream alters surface tension and affects k. A nonwettable material destroys bubble stability. When long delay coils are needed, polythene coils are better than glass and Tygon is slightly worse than glass (884). There are no quantitative data available on the effect of tube bore but it is evident that, within limits available on existing equipment, reduction in tube bore reduces σ_D.

2. Square-Wave Regeneration

Since the mathematical nature of the σ_D deformation is defined, it is possible to compute the original input square wave from the deformation left after removal of the exponential component. In practice, such computation has proven to be unhelpful because intrinsic machine noise accumulates to an unacceptable extent. It has been shown that the simple exponential regeneration procedure is unsuited to the reversal of σ_D deformation (70a) and in general any such attempt leads to overshot of the next plateau position.

3. Design Changes

Design changes from AutoAnalyzer I to AutoAnalyzer II equipment have resulted in marked reduction of σ_D deformation, and this process has been extended to its logical conclusion in the SMAC system. The increase in bubble frequency has required precise metering of bubble interval to remove noise due to bubble irregularity (253).

VII. MAXIMUM SAMPLING RATE

If exponential deformation is eliminated, σ_D becomes the limiting factor in sampling rate. Most AutoAnalyzer II methods have σ_D values of about 2.5 sec. With $t_s = 4.0\sigma_D$ and $t_w = 1.0\sigma_D$, peak heights of 95% of plateau result with 0.1% interaction and shoulder peaking when adjacent peak ratios exceed 4.5. In a noise-free system, sampling rates of about 300 per hour would be possible. In practice, both curve regeneration circuits (886) and electronic bubble-gating circuits (623a) require smoothing filters to eliminate low-frequency noise. Such smoothing imposes a small deformation of its own. With either of these approaches, sampling rates of 150 per hour are readily attainable with good quality output. With AutoAnalyzer I systems, similar results are obtained at sampling rates of 100 per hour (166a). At these sampling rates, the mechanical cams used for sampler timing should be replaced by electronic timers in order to avoid imprecision due to cam irregularities (926). Modern samplers have such a rapid vertical movement of the sample probe into and out of sample cups that, even at high sampling rates, the depth of liquid in the sample cups does not measurably influence the true sampling time in the way that it did in earlier samplers (835).

VIII. CONCLUSION

Understanding of the nature of deformation effects in continuous flow streams now enables subjective design and evaluation to be replaced by objective measures of kinetic characteristics. The relevant deformation constants are easily measured and should be used together with a measure of noise level at baseline and maximum absorbance for description, comparison, and evaluation purposes. They provide an effective means of monitoring performance and give some indication of the likely location of system deterioration (832). Application of these principles has led to improvements in operating conditions and to electronic compensating devices which together result in better precision and faster sampling rates.

REFERENCES

1. M. I. Abdullah, in <u>Automation in Analytical Chemistry, Technicon Symposia 1967</u>, Vol. II, Mediad, Inc., White Plains, N.Y., 1968, pp. 401-405.

2. M. I. Abdullah and J. P. Riley, in <u>Automation in Analytical Chemistry, Technicon Symposia 1966</u>, Vol. II, Mediad, Inc., White Plains, N.Y., 1967, pp. 85-95.

3. M. H. Adelman, in <u>Automation in Analytical Chemistry, Technicon Symposia 1966</u>, Vol. I, Mediad, Inc., White Plains, N.Y., 1967, pp. 552-556.

4. M. H. Adelman and R. L. Pellissier, in <u>Automation in Analytical Chemistry, Technicon Symposia 1965</u>, Mediad, Inc., New York, 1966, pp. 183-185.

5. S. L. Adler, Sr., and A. Tse, in <u>Advances in Automated Analysis, Technicon International Congress 1970</u>, Vol. I, Thurman Associates, Miami, 1971, pp. 421-424.

6. B. K. Afghan, P. D. Goulden, and J. F. Ryan, in <u>Advances in Automated Analysis, Technicon International Congress 1970</u>, Vol. II, Thurman Associates, Miami, 1971, pp. 291-297.

7. B. K. Afghan, P. D. Goulden, and J. F. Ryan, <u>Anal. Chem.</u>, <u>44</u>, 354-359 (1972).

8. A. Agren and E. R. Garrett, <u>Acta Pharm. Suecica</u>, <u>4</u>, 1-12 (1967).

9. J. N. Ahuja, A. Kaplan, and P. Van Dreal, <u>Clin. Chem.</u>, <u>14</u>, 664-674 (1968).

10. S. Ahuja and C. Spitzer, in <u>Advances in Automated Analysis, Technicon International Congress 1970</u>, Vol. II, Thurman Associates, Miami, 1971, pp. 227-232.

11. S. Ahuja, C. Spitzer, and F. R. Brofazi, in Automation in Analytical Chemistry, Technicon Symposia 1967, Vol. I, Mediad, Inc., White Plains, N.Y., 1968, pp. 439-442.

12. S. Ahuja, C. Spitzer, and F. R. Brofazi, in Automation in Analytical Chemistry, Technicon Symposia 1967, Vol. I, Mediad, Inc., White Plains, N.Y., 1968, pp. 467-470.

13. S. Ahuja, C. Spitzer, and F. R. Brofazi, J. Pharm. Sci., 59, 1833-1835 (1970).

14. L. L. Alber, M. W. Overton, and D. E. Smith, J. Ass. Offic. Anal. Chem., 56, 659-666 (1973).

15. A. M. Albisser and B. S. Leibel, in Advances in Automated Analysis, Technicon International Congress 1970, Vol. I, Thurman Associates, Miami, 1971, pp. 535-538.

16. B. E. Albright, in Automation in Analytical Chemistry, Technicon Symposia 1967, Vol. I, Mediad, Inc., White Plains, N.Y., 1968, pp. 443-446.

17. E. Allen, in Automation in Analytical Chemistry, Technicon Symposia 1965, Mediad, Inc., New York, 1966, pp. 328-330.

18. E. Allen, Clin. Chem., 12, 376-378 (1966).

19. N. L. Alpert, Clin. Chem., 15, 1198-1209 (1969).

20. I. Alsos, V. Holm, Y. Torud, and T. Waaler, Medd. Nor. Farm. Selsk., 31, 115-124 (1969).

21. I. Alsos, V. Holm, T. Willadssen, and T. Waaler, Medd. Nor. Farm. Selsk., 31, 11-16 (1969).

22. E. Amador, Clin. Chem., 18, 164 (1972).

23. E. Amador, R. L. Cechner, and J. J. Barklow, Clin. Chem., 18, 668-671 (1972).

24. E. Amador and J. Urban, Clin. Chem., 18, 601-604 (1972).

25. J. A. Ambrose, in Advances in Automated Analysis, Technicon International Congress 1969, Vol. I, Mediad, Inc., White Plains, N.Y., 1970, pp. 25-32.

26. J. A. Ambrose, in Advances in Automated Analysis, Technicon International Congress 1970, Vol. I, Thurman Associates, Miami, 1971, pp. 517-521.

27. J. A. Ambrose, Ann. N.Y. Acad. Sci., 196, 295-303 (1972).

28. J. A. Ambrose, C. Eigel, and C. Smith, in Advances in Automated Analysis, Technicon International Congress 1969, Vol. I, Mediad, Inc., White Plains, N.Y., 1970, pp. 91-93.

29. J. A. Ambrose, C. Ross, and F. Whitfield, in Automation in Analytical Chemistry, Technicon Symposia 1967, Vol. I, Mediad, Inc., White Plains, N.Y., 1968, pp. 13-19.

30. N. G. Anderson, Anal. Chem., 33, 970-971 (1961).

31. R. A. Anderson, C. Perrizo, and S. A. Fusari, Ann. N.Y. Acad. Sci., 153, 471-485 (1968).

32. R. A. Anderson, C. Perrizo, and S. A. Fusari, in Automation in Analytical Chemistry, Technicon Symposia 1966, Vol. I, Mediad, Inc., White Plains, N.Y., 1967, pp. 267-272.

33. J. S. Annino, Amer. J. Clin. Pathol., 45, 512-514 (1966).

34. J. S. Annino, Tech. Bull. Regist. Med. Technol. Amer. Soc. Clin. Pathol., 36, 68-70 (1966).

35. Anon., in Handbook of Clinical Laboratory Data (H. C. Damm and J. W. King, eds.), Chemical Rubber Co., Cleveland, 1965, pp. 217-226.

36. Anon., J. Ass. Offic. Anal. Chem., 55, 422-424 (1972).

36a. Anon., J. Clin. Pathol., 22, 278-284 (1969).

37. C. L. Anstiss, S. Green, and W. H. Fishman, Clin. Chim. Acta, 33, 279-286 (1971).

38. A. Antonis, J. Lipid Res., 6, 307-312 (1965).

39. A. Antonis, M. Clark, and T. R. E. Pilkington, J. Lab. Clin. Med., 68, 340-356 (1966).

40. J. L. Arbogast, Amer. J. Clin. Pathol., 38, 176 (1962).

41. N. P. Archer, E. C. Pierce, and R. C. Anderson, Amer. J. Clin. Pathol., 46, 89-91 (1966).

42. R. A. Arend, J. E. Leklem, and R. R. Brown, in Advances in Automated Analysis, Technicon International Congress 1969, Vol. II, Mediad, Inc., White Plains, N.Y., 1970, pp. 195-199.

43. W. G. Armstrong, in Automation in Analytical Chemistry, Technicon Symposia 1967, Vol. I, Mediad, Inc., White Plains, N.Y., 1968, pp. 295-299.

44. T. W. L. Atkinson, Chem. Process. (London), 28, April (1965).

45. H. B. Auerbach and R. R. Bartchy, in Automation in Analytical Chemistry, Technicon Symposia 1966, Vol. I, Mediad, Inc., White Plains, N.Y., 1967, pp. 222-223.

46. Automating Manual Methods using Technicon AutoAnalyzer II System Techniques, Manual TN 1-0170-01, Technicon Instruments Corporation, Tarrytown, N.Y., 1972.

47. Automation and Data Processing in the Clinical Laboratory (G. M. Brittin and M. Werner, eds.), Thomas, Springfield, Ill., 1970.

48. F. Avanzini, D. Magnanelli, and V. Boffi, in Automation in Analytical Chemistry, Technicon Symposia 1967, Vol. II, Mediad, Inc., White Plains, N.Y., 1968, pp. 285-290.

49. H. Axelsson, B. Ekman, and D. Knutsson, in Automation in Analytical Chemistry, Technicon Symposia 1965, Mediad, Inc., New York, 1966, pp. 603-608.

50. C. W. Ayers, in Automation in Analytical Chemistry, Technicon Symposia 1966, Vol. II, Mediad, Inc., White Plains, N.Y., 1967, pp. 107-110.

51. A. L. Babson and N. M. Kleinman, Clin. Chem., 13, 163-166 (1967).

52. K. D. Bagshawe, F. W. Harris, and A. H. Orr, in Automation in Analytical Chemistry, Technicon Symposia 1967, Vol. II, Mediad, Inc., White Plains, N.Y., 1968, pp. 53-56.

53. B. W. Bailey and F. C. Lo, Anal. Chem., 43, 1525-1526 (1971).

54. J. D. Baird, M. W. Black, and D. E. Faulkner, in Automation in Analytical Chemistry, Technicon Symposia 1967, Vol. II, Mediad, Inc., White Plains, N.Y., 1968, pp. 105-110.

55. J. D. Baird, M. W. Black, and D. E. Faulkner, J. Clin. Pathol., 20, 905-909 (1967).

56. R. K. Baker, A. Posner, and H. Francis, in Advances in Automated Analysis, Technicon International Congress 1969, Vol. I, Mediad, Inc., White Plains, N.Y., 1970, pp. 175-177.

57. E. A. Balazs, K. O. Berntsen, J. Karossa, and D. A. Swann, Anal. Biochem., 12, 559-564 (1965).

58. F. J. Bano and R. J. Crossland, Analyst (London), 97, 823-827 (1972).

59. S. Barabas, in Automation in Analytical Chemistry, Technicon Symposia 1966, Vol. I, Mediad, Inc., White Plains, N.Y., 1967, pp. 617-621.

60. S. Barabas and S. G. Lea, Anal. Chem., 37, 1132-1135 (1965).

61. S. Barabas and S. G. Lea, in Automation in Analytical Chemistry, Technicon Symposia 1965, Mediad, Inc., New York, 1966, pp. 164-168.

62. G. W. Barber, in Automation in Analytical Chemistry, Technicon Symposia 1966, Vol. I, Mediad, Inc., White Plains, N.Y., 1967, pp. 401-405.

63. H. Barrera, K. S. Chio, and A. L. Tappel, Anal. Biochem., 29, 515-525 (1969).

64. E. T. Bartley, Jr., and M. D. Poulik, in Automation in Analytical Chemistry, Technicon Symposia 1966, Vol. I, Mediad, Inc., White Plains, N.Y., 1967, pp. 383-387.

65. W. D. Basson and R. G. Bohmer, Analyst (London), 97, 266-270 (1972).

66. E. H. Baum, Ann. N.Y. Acad. Sci., 87, 894-903 (1960).

67. A. N. Baumann and H. H. Roberts, in Automation in Analytical Chemistry, Technicon Symposia 1967, Vol. I, Mediad, Inc., White Plains, N.Y., 1968, pp. 257-260.

68. C. Beck and A. L. Tappel, Anal. Biochem., 21, 208-218 (1967).

69. R. D. Begg, Anal. Chem., 43, 854-857 (1971).

70. R. D. Begg, Anal. Chem., 44, 631-632 (1972).

70a. R. D. Begg, Anal. Chem., 46, 171-173 (1974).

71. J. L. Bell, M. Collier, and A. M. Hartmann, J. Clin. Pathol., 17, 301-303 (1964).

72. A. Bennet, D. Gartelmann, J. I. Mason, and J. A. Owen, Clin. Chim. Acta, 29, 161-180 (1970).

73. C. Benz and R. M. Kelley, in Advances in Automated Analysis, Technicon International Congress 1969, Vol. II, Mediad, Inc., White Plains, N.Y., 1970, pp. 245-249.

74. C. Benz and L. M. Paixao, in Automation in Analytical Chemistry, Technicon Symposia 1967, Vol. I, Mediad, Inc., White Plains, N.Y., 1968, pp. 223-225.

75. M. Bernhard and G. Macchi, in Automation in Analytical Chemistry, Technicon Symposia 1965, Mediad, Inc., New York, 1966, pp. 255-259.

76. M. Bernhard, E. Torti, M. Ghibaudo, G. Rossi, and A. Bruschi, in Automation in Analytical Chemistry, Technicon Symposia 1967, Vol. II, Mediad, Inc., White Plains, N.Y., 1968, pp. 391-394.

77. M. Bernhard, E. Torti, and G. Rossi, in Automation in Analytical Chemistry, Technicon Symposia 1967, Vol. II, Mediad, Inc., White Plains, N.Y., 1968, pp. 395-400.

78. C. T. Berry and R. J. Crossland, Analyst (London), 95, 291-295 (1970).

79. M. N. Berry, in Automation and Data Processing in the Clinical Laboratory (G. M. Brittin and M. Werner, eds.), Thomas, Springfield, Ill., 1970, pp. 151-158.

80. M. N. Berry and A. K. Walli, in Automation in Analytical Chemistry, Technicon Symposia 1966, Vol. II, Mediad, Inc., White Plains, N.Y., 1967, pp. 389-393.

81. H. W. Bertram, M. W. Lerner, G. J. Petretic, E. S. Roszkowski, and C. J. Rodden, Anal. Chem., 30, 354-359 (1958).

82. V. G. Bethune, M. Fleisher, and M. K. Schwartz, Clin. Chem., 18, 1524-1528 (1972).

83. W. F. Beyer, in Automation in Analytical Chemistry, Technicon Symposia 1966, Vol. I, Mediad, Inc., White Plains, N.Y., 1967, pp. 259-262.

84. W. F. Beyer, J. Pharm. Sci., 55, 200-203 (1966).

85. W. F. Beyer, J. Pharm. Sci., 55, 622-625 (1966).

86. W. F. Beyer, J. Pharm. Sci., 56, 526-529 (1967).

87. W. F. Beyer, J. Pharm. Sci., 57, 1415-1418 (1968).

88. W. Beyer and W. Morozowich, Ann. N.Y. Acad. Sci., 153, 393-402 (1968).

89. W. F. Beyer and E. W. Smith, in Advances in Automated Analysis, Technicon International Congress 1969, Vol. II, Mediad, Inc., White Plains, N.Y., 1970, pp. 217-223.

90. W. F. Beyer and E. W. Smith, J. Pharm. Sci., 59, 248-253 (1970).

91. R. W. Bide, Anal. Biochem., 29, 393-404 (1969).

92. R. W. Bide and W. J. Dorward, in Advances in Automated Analysis, Technicon International Congress 1970, Vol. II, Thurman Associates, Miami, 1971, pp. 423-427.

93. B. A. Bidlingmeyer and S. N. Deming, Anal. Biochem., 43, 582-587 (1971).

94. A. Bieder and P. Brunel, Ann. N.Y. Acad. Sci., 130, 627-633 (1965).

95. C. G. Bird and J. A. Owen, Clin. Chim. Acta, 24, 305-307 (1969).

96. D. J. Birkett, R. A. J. Conyers, F. C. Neale, S. Posen, and J. Brudenell-Woods, Arch. Biochem. Biophys., 121, 470-479 (1967).

97. D. L. Bittner and J. Manning, in Automation in Analytical Chemistry, Technicon Symposia 1966, Vol. I, Mediad, Inc., White Plains, N.Y., 1967, pp. 33-36.

98. F. Bjorksten, Clin. Chim. Acta, 40, 143-152 (1972).

99. D. J. Blackmore, A. S. Curry, T. S. Hayes, and E. R. Rutter, Clin. Chem., 17, 896-902 (1971).

100. L. G. Blackwell, in Advances in Automated Analysis, Technicon International Congress 1970, Vol. II, Thurman Associates, Miami, 1971, pp. 341-345.

101. W. J. Blaedel and G. P. Hicks, in Advances in Analytical Chemistry and Instrumentation (C. N. Reilley, ed.), Vol. 3, Wiley (Interscience), New York, 1964, pp. 105-142.

102. W. J. Blaedel and G. P. Hicks, Anal. Biochem., 4, 476-488 (1962).

103. W. J. Blaedel and G. P. Hicks, Anal. Chem., 34, 388-394 (1962).

104. W. J. Blaedel and R. H. Laessig, in Advances in Analytical Chemistry and Instrumentation (C. N. Reilley and F. W. McLafferty, eds.), Vol. 5, Wiley (Interscience), New York, 1966, pp. 69-168.

105. W. J. Blaedel and E. D. Olsen, Anal. Chem., 32, 789-791 (1960).

106. W. J. Blaedel and C. L. Olson, J. Chem. Educ., 40, A549-A558 (1963).

107. W. J. Blaedel and J. H. Strohl, Anal. Chem., 33, 1631-1632 (1961).

108. W. J. Blaedel and J. H. Strohl, Anal. Chem., 36, 445-446 (1964).

109. W. J. Blaedel and J. H. Strohl, Anal. Chem., 36, 1245-1251 (1964).

110. W. J. Blaedel and J. W. Todd, Anal. Chem., 30, 1821-1825 (1958).

111. W. J. Blaedel and J. W. Todd, Anal. Chem., 33, 205-207 (1961).

112. M. A. Blaivas, in Automation in Analytical Chemistry, Technicon Symposia 1965, Mediad, Inc., New York, 1966, pp. 452-454.

113. M. A. Blaivas and A. H. Mencz, in Automation in Analytical Chemistry, Technicon Symposia 1966, Vol. I, Mediad, Inc., White Plains, N.Y., 1967, pp. 368-372.

114. M. A. Blaivas and A. H. Mencz, in Automation in Analytical Chemistry, Technicon Symposia 1967, Vol. I, Mediad, Inc., White Plains, N.Y., 1968, pp. 133-136.

115. R. G. Blezard and J. A. Fifield, in Advances in Automated Analysis, Technicon International Congress 1969, Vol. II, Mediad, Inc., White Plains, N.Y., 1970, pp. 283-295.

116. R. J. Block and R. H. Mandl, Biochem. J., 81, 37P-38P (1961).

117. A. M. Bold, J. Clin. Pathol., 19, 625-628 (1966).

118. A. M. Bold, R. J. Hurst, and R. R. McSwiney, J. Clin. Pathol., 18, 240-243 (1965).

119. J. Bomstein and W. G. Evans, Anal. Chem., 37, 576-578 (1965).

120. J. Bomstein, W. G. Evans, R. L. McLaughlin, J. P. Messerly, and J. M. Shepp, Anal. Chem., 36, 1149-1150 (1964).

121. J. Bomstein, W. G. Evans, and J. M. Shepp, Ann. N.Y. Acad. Sci., 130, 589-595 (1965).

122. J. Bomstein, J. M. Shepp, S. T. Dawson, and W. J. Blaedel, J. Pharm. Sci., 55, 94-98 (1966).

123. S. Bonk and L. N. Scharfe, in Automation in Analytical Chemistry, Technicon Symposia 1966, Vol. I, Mediad, Inc., White Plains, N.Y., 1967, pp. 638-642.

124. R. J. Boucher, K. D. Luke, and H. H. Wood, in Advances in Automated Analysis, Technicon International Congress 1969, Vol. II, Mediad, Inc., White Plains, N.Y., 1970, pp. 39-44.

125. J. Bourdillon and R. E. Vanderlinde, Pub. Health Rep., 81, 991-995 (1966).

126. D. W. Bradley and A. L. Tappel, Anal. Biochem., 33, 400-413 (1970).

127. J. S. Bradshaw and C. W. Spanis, in Advances in Automated Analysis, Technicon International Congress 1970, Vol. I, Thurman Associates, Miami, 1971, pp. 17-26.

128. J. S. Bradshaw and C. W. Spanis, in Advances in Automated Analysis, Technicon International Congress 1970, Vol. II, Thurman Associates, Miami, 1971, pp. 17-26.

129. A. L. Brandon, in Automation in Analytical Chemistry, Technicon Symposia 1966, Vol. I, Mediad, Inc., White Plains, N.Y., 1967, pp. 584-590.

130. M. Brandstein, P. Glass, A. Castellano, and C. Mezzacappa, Amer. J. Clin. Pathol., 42, 218-220 (1964).

131. G. Brecher and H. F. Loken, Amer. J. Clin. Pathol., 55, 527-540 (1971).

132. M. Breen and R. T. Marshall, J. Lab. Clin. Med., 68, 701-712 (1966).

133. P. G. Brewer, K. M. Chan, and J. P. Riley, in Automation in Analytical Chemistry, Technicon Symposia 1965, Mediad, Inc., New York, 1966, pp. 308-314.

134. G. B. Briscoe, B. G. Cooksey, M. Williams, and J. Ruzicka, in Automation in Analytical Chemistry, Technicon Symposia 1967, Vol. II, Mediad, Inc., White Plains, N.Y., 1968, pp. 309-314.

135. R. D. Britt, Jr., Anal. Chem., 34, 1728-1731 (1962).

136. A. Brodie and A. Hanok, Clin. Chem., 16, 65 (1970).

137. L. Brooks and H. G. Olken, Clin. Chem., 11, 748-762 (1965).

138. P. M. G. Broughton, M. A. Buttolph, A. H. Gowenlock, D. W. Neill, and R. G. Skentelbery, J. Clin. Pathol., 22, 278-284 (1969).

139. E. V. Browett and R. Moss, Analyst (London), 90, 715-726 (1965).

140. E. V. Browett and R. Moss, in Automation in Analytical Chemistry, Technicon Symposia 1965, Mediad, Inc., New York, 1966, pp. 375-382.

141. D. J. Brown, J. E. Nohlgren, and A. A. Oyama, in Automation in Analytical Chemistry, Technicon Symposia 1966, Vol. I, Mediad, Inc., White Plains, N.Y., 1967, pp. 304-308.

142. E. O. Brown, in Practical Automation for the Clinical Laboratory (W. L. White, M. M. Erickson, and S. C. Stevens, eds.), C. V. Mosby Co., St. Louis, 1968, pp. 291-308.

143. G. M. Brown, A. Zachwieja, and H. C. Stancer, Clin. Chim. Acta, 14, 386-390 (1966).

144. H. H. Brown, Clin. Chem., 14, 967-978 (1968).

145. H. H. Brown and M. R. Ebner, Clin. Chem., 13, 847-854 (1967).

146. H. H. Brown, M. C. Rhindress, and R. E. Griswold, Clin. Chem., 17, 92-96 (1971).

147. M. E. Brown, B. Blaney, and A. Labuckas, in Automation in Analytical Chemistry, Technicon Symposia 1967, Vol. I, Mediad, Inc., White Plains, N.Y., 1968, pp. 35-39.

148. M. C. Brummel, H. E. Mayer, Jr., and R. Montgomery, Anal. Biochem., 33, 16-27 (1970).

149. R. Bryant, F. J. Burger, and F. B. Trenk, J. Pharm. Sci., 60, 1721-1722 (1971).

150. E. B. Buchanan, Jr., and J. R. Bacon, Anal. Chem., 39, 615-620 (1967).

151. N. R. M. Buist and J. J. Strandholm, in Advances in Automated Analysis, Technicon International Congress 1969, Vol. I, Mediad, Inc., White Plains, N.Y., 1970, pp. 321-326.

152. D. A. Burns, in Automation in Analytical Chemistry, Technicon Symposia 1965, Mediad, Inc., New York, 1966, pp. 193-196.

153. D. A. Burns, in Automation in Analytical Chemistry, Technicon Symposia 1966, Vol. I, Mediad, Inc., White Plains, N.Y., 1967, pp. 93-99.

154. D. A. Burns and W. C. Bass, in Advances in Automated Analysis, Technicon International Congress 1969, Vol. I, Mediad, Inc., White Plains, N.Y., 1970, pp. 95-99.

155. D. A. Burns and G. D. Hansen, Ann. N.Y. Acad. Sci., 153, 541-558 (1968).

156. J. R. Burt, Anal. Biochem., 9, 293-302 (1964).

157. B. Cacciapuotti, in Advances in Automated Analysis, Technicon International Congress 1970, Vol. I, Thurman Associates, Miami, 1971, pp. 325-329.

158. N. G. Cadavid and A. C. Paladini, Anal. Biochem., 9, 170-174 (1964).

159. M. C. Cadmus and G. W. Strandberg, Anal. Biochem., 26, 484-487 (1968).

160. J. D. Caisey and B. D. Riordan, Analyst (London), 98, 126-132 (1973).

161. L. J. Cali, J. M. Konieczny, and J. D. DeMarco, in Automation in Analytical Chemistry, Technicon Symposia 1967, Vol. I, Mediad, Inc., White Plains, N.Y., 1968, pp. 451-455.

162. D. G. Campbell, Clin. Chim. Acta, 33, 473-474 (1971).

163. D. G. Campbell and G. Gardner, Clin. Chim. Acta, 32, 153-157 (1971).

164. T. E. Cantor, in Automation in Analytical Chemistry, Technicon Symposia 1966, Vol. I, Mediad, Inc., White Plains, N.Y., 1967, pp. 514-515.

165. F. Cappellina and L. Cappelletti, in Automation in Analytical Chemistry, Technicon Symposia 1967, Vol. II, Mediad, Inc., White Plains, N.Y., 1968, pp. 295-301.

166. R. D. Capps, II, J. G. Batsakis, R. O. Briere, and R. R. L. Calam, Clin. Chem., 12, 406-413 (1966).

166a. J. E. Carlyle, A. S. McLelland, and A. Fleck, Clin. Chim. Acta, 46, 235-241 (1973).

167. R. F. Carroll, in Fourth National Pulp and Paper Instrumentation Symposium, Memphis, Tennessee, May 8-10, 1963, "Digital Computer Control of a Chemical Process."

168. M. E. Carruthers, J. Clin. Pathol., 23, 450-451 (1970).

169. J. M. Carter and G. Nickless, Analyst (London), 95, 148-152 (1970).

170. E. W. Catanzaro, in 1964 Technicon International Symposium, Automation in Analytical Chemistry, New York, "Some Fundamental Aspects of the Technicon Automatic Kjeldahl Analyzer ... Digestion Techniques."

171. E. W. Catanzaro, D. R. Grady, N. S. Zaleiko, and R. A. Olsen, in Automation in Analytical Chemistry, Technicon Symposia 1967, Vol. I, Mediad, Inc., White Plains, N.Y., 1968, pp. 357-361.

172. G. N. Catravas, Anal. Chem., 36, 1146-1148 (1964).

173. G. N. Catravas, in Automation in Analytical Chemistry, Technicon Symposia 1966, Vol. I, Mediad, Inc., White Plains, N.Y., 1967, pp. 397-400.

174. G. N. Catravas and E. D. Lash, in Automation in Analytical Chemistry, Technicon Symposia 1966, Vol. I, Mediad, Inc., White Plains, N.Y., 1967, pp. 445-446.

175. L. Cavatorta, R. Rangone, C. Calanni, and M. Casa, in Advances in Automated Analysis, Technicon International Congress 1969, Vol. II, Mediad, Inc., White Plains, N.Y., 1970, pp. 261-264.

176. B. E. Cham, Clin. Chim. Acta, 37, 5-14 (1972).

177. A. L. Chaney, in Automation in Analytical Chemistry, Technicon Symposia 1967, Vol. I, Mediad, Inc., White Plains, N.Y., 1968, pp. 115-117.

178. A. L. Chasson, Clin. Chem., 17, 351-352 (1971).

179. A. L. Chasson, J. Jackson, and M. West, in Automation in Analytical Chemistry, Technicon Symposia 1966, Vol. I, Mediad, Inc., White Plains, N.Y., 1967, pp. 315-320.

180. C. S. Cheek, in Automation in Analytical Chemistry, Technicon Symposia 1965, Mediad, Inc., New York, 1966, pp. 458-462.

181. R. H. Cherry, Ann. N.Y. Acad. Sci., 91, 850-872 (1961).

182. A. Chersi, P. G. Pietta, and A. Corbellini, in Automation in Analytical Chemistry, Technicon Symposia 1967, Vol. II, Mediad, Inc., White Plains, N.Y., 1968, pp. 223-226.

183. J. Chevron, L. Terrier, L. Foissac, and J. J. Verniere, in
     Automation in Analytical Chemistry, Technicon Symposia 1966, Vol. I,
     Mediad, Inc., White Plains, N.Y., 1967, pp. 627-632.

184. K. J. Child and J. D. Caisey, in Automation in Analytical Chemistry,
     Technicon Symposia 1966, Vol. II, Mediad, Inc., White Plains, N.Y.,
     1967, pp. 195-201.

185. H. L. Christopherson, J. Ass. Offic. Anal. Chem., 52, 93-101 (1969).

186. D. D. Clarke and R. Nicklas, in Automation in Analytical Chemistry,
     Technicon Symposia 1966, Vol. I, Mediad, Inc., White Plains, N.Y.,
     1967, pp. 465-468.

187. J. Clements and J. F. Marten, in Automation in Analytical Chemistry,
     Technicon Symposia 1966, Vol. I, Mediad, Inc., White Plains, N.Y.,
     1967, pp. 183-186.

188. A. Cohen, in Automation in Analytical Chemistry, Technicon Symposia
     1966, Vol. II, Mediad, Inc., White Plains, N.Y., 1967, pp. 341-346.

189. A. Cohen, J. Clin. Pathol., 19, 589-594 (1966).

190. M. F. Collen, L. Morgenstern, and Y. R. Rao, in 1964 Technicon
     International Symposium, New York, "Automated Chemical Analysis
     in Multiphasic Screening."

191. D. U. Conaill and G. G. Muir, in Automation in Analytical Chemistry,
     Technicon Symposia 1967, Vol. II, Mediad, Inc., White Plains, N.Y.,
     1968, pp. 137-142.

192. D. U. Conaill and G. G. Muir, Clin. Chem., 14, 1010-1022 (1968).

193. N. Conca and H. J. Pazdera, Ann. N.Y. Acad. Sci., 130, 596-601
     (1965).

194. W. J. Constandse, Ann. N.Y. Acad. Sci., 153, 609-619 (1968).

195. W. J. Constandse, in Automation in Analytical Chemistry, Technicon
     Symposia 1966, Vol. I, Mediad, Inc., White Plains, N.Y., 1967,
     pp. 357-363.

196. W. D. Conway and E. J. Lethco, Anal. Chem., 34, 447-448 (1962).

197. W. Cool and Y. Annokkee, in Advances in Automated Analysis,
     Technicon International Congress 1970, Vol. I, Thurman Associates,
     Miami, 1971, pp. 607-618.

198. A. R. Coote, I. W. Duedall, and R. S. Hiltz, in Advances in Automated
     Analysis, Technicon International Congress 1970, Vol. II, Thurman
     Associates, Miami, 1971, pp. 347-351.

199. E. Cotlove, Clin. Chem., 11, 816 (1965).

200. C. R. Craig, A. J. Azzaro, B. K. Frame, and W. A. Hunt, in Advances in Automated Analysis, Technicon International Congress 1969, Vol. II, Mediad, Inc., White Plains, N.Y., 1970, pp. 189-193.

201. R. M. Creamer, Anal. Chem., 29, 1722-1723 (1957).

202. R. J. Creno, R. E. Wenk, and P. Bohlig, Amer. J. Clin. Pathol., 54, 828-832 (1970).

203. R. F. Crerand, H. Hochstrasser, S. Morgenstern, Y. R. Yoon, and R. Varady, in Advances in Automated Analysis, Technicon International Congress 1970, Vol. I, Thurman Associates, Miami, 1971, pp. 37-40.

204. A. M. Crestfield, Anal. Chem., 35, 1762-1763 (1963).

205. O. B. Crofford and W. W. Lacy, J. Lab. Clin. Med., 61, 708-712 (1963).

206. L. V. Crowley, Clin. Chim. Acta, 11, 386-388 (1965).

207. L. F. Cullen, L. J. Heckman, and G. J. Papariello, in Advances in Automated Analysis, Technicon International Congress 1969, Vol. II, Mediad, Inc., White Plains, N.Y., 1970, pp. 233-238.

208. L. F. Cullen, L. J. Heckman, and G. J. Papariello, J. Pharm. Sci., 58, 1537-1539 (1969).

209. L. F. Cullen, D. L. Packman, and G. J. Papariello, in Advances in Automated Analysis, Technicon International Congress 1969, Vol. II, Mediad, Inc., White Plains, N.Y., 1970, pp. 213-216.

210. L. F. Cullen, D. L. Packman, and G. J. Papariello, J. Pharm. Sci., 59, 697-699 (1970).

211. L. F. Cullen, J. G. Rutgers, P. A. Lucchesi, and G. J. Papariello, J. Pharm. Sci., 57, 1857-1864 (1968).

212. B. A. Cunningham, P. D. Gottlieb, W. H. Konigsberg, and G. M. Edelman, Biochemistry, 7, 1983-1995 (1968).

213. F. W. Czech and T. P. Hrycyshyn, in Automation in Analytical Chemistry, Technicon Symposia 1966, Vol. I, Mediad, Inc., White Plains, N.Y., 1967, pp. 163-166.

214. F. R. Dalal and S. Winsten, Clin. Chem., 16, 969 (1970).

215. C. Dalton and C. Kowalski, Clin. Chem., 13, 744-751 (1967).

216. R. S. Danchik and R. T. Oliver, Anal. Chem., 42, 798-801 (1970).

217. J. Davidson, J. Mathieson, and A. W. Boyne, Analyst (London), 95, 181-193 (1970).

218. A. W. Davies and K. Taylor, in Automation in Analytical Chemistry, Technicon Symposia 1965, Mediad, Inc., New York, 1966, pp. 294-300.

219. W. C. Davies, A. J. Goudie, and A. Khoury, in Automation in Analytical Chemistry, Technicon Symposia 1967, Vol. I, Mediad, Inc., White Plains, N.Y., 1968, pp. 457-459.

220. H. A. Davis, R. E. Sterling, A. A. Wilcox, and W. E. Waters, Clin. Chem., 12, 428-431 (1966).

221. R. E. Davis, D. J. Nicol, and A. Kelly, J. Clin. Pathol., 23, 47-53 (1970).

222. C. A. Dawes and J. A. Tetlow, in Automation in Analytical Chemistry, Technicon Symposia 1965, Mediad, Inc., New York, 1966, pp. 232-236.

223. R. Delaney, Anal. Biochem., 46, 413-420 (1972).

224. H. T. Delves and P. Vinter, J. Clin. Pathol., 19, 504-509 (1966).

225. T. F. Demmitt, in Automation in Analytical Chemistry, Technicon Symposia 1965, Mediad, Inc., New York, 1966, pp. 204-206.

226. R. Dewart, F. Naudts, and W. Lhoest, Ann. N.Y. Acad. Sci., 130, 686-696 (1965).

227. J. P. Dieu, Clin. Chem., 17, 1183-1185 (1971).

228. A. C. Docherty, Lab. Pract., 18, (1), 47-51 (1969).

229. A. C. Docherty, Proc. Fert. Soc., 104, 45-61 (1968).

230. M. L. Dow, J. Ass. Offic. Anal. Chem., 56, 647-652 (1973).

231. N. E. Dowd, A. M. Killard, H. J. Pazdera, and A. Ferrari, Ann. N.Y. Acad. Sci., 130, 558-567 (1965).

232. N. E. Dowd, R. J. Raffa, and H. J. Pazdera, in Automation in Analytical Chemistry, Technicon Symposia 1966, Vol. I, Mediad, Inc., White Plains, N.Y., 1967, pp. 263-266.

233. L. Dufresne and H. J. Gitelman, Anal. Biochem., 37, 402-408 (1970).

234. R. E. Duncombe and W. H. C. Shaw, in Automation in Analytical Chemistry, Technicon Symposia 1966, Vol. II, Mediad, Inc., White Plains, N.Y., 1967, pp. 15-18.

235. C. W. Easley, B. P. Godwin, and H. Kitchen, in Automation in Analytical Chemistry, Technicon Symposia 1966, Vol. I, Mediad, Inc., White Plains, N.Y., 1967, pp. 427-434.

236. I. Eckman, J. B. Robbins, C. J. A. Van den Hamer, J. Lentz, and I. H. Scheinberg, Clin. Chem., 16, 558-561 (1970).

237. D. L. Eichler, in Advances in Automated Analysis, Technicon International Congress 1969, Vol. II, Mediad, Inc., White Plains, N.Y., 1970, pp. 51-59.

238. H. P. Eiduson and W. M. Hoffman, J. Ass. Offic. Anal. Chem., 50, 1021-1023 (1967).

239. L. Ek, J. Fernandez, and L. C. Leeper, in Automation in Analytical Chemistry, Technicon Symposia 1967, Vol. I, Mediad, Inc., White Plains, N.Y., 1968, pp. 477-482.

240. J. F. Elam and R. E. Palmer, Amer. J. Clin. Pathol., 46, 65-68 (1966).

241. T. Ellerington and D. C. Ferguson, in Automation in Analytical Chemistry, Technicon Symposia 1967, Vol. II, Mediad, Inc., White Plains, N.Y., 1968, pp. 417-421.

242. F. W. Ellis and J. B. Hill, Clin. Chem., 15, 91-101 (1969).

243. J. H. Ellis, J. Ass. Offic. Anal. Chem., 49, 870-872 (1966).

244. E. A. Epps, Jr., and H. C. Austin, Jr., J. Ass. Offic. Anal. Chem., 50, 981-982 (1967).

245. G. Ertingshausen, H. J. Adler, and A. S. Reichler, J. Chromatogr., 42, 355-366 (1969).

246. G. Ertingshausen, H. J. Adler, A. S. Reichler, and N. Kinnard, in Advances in Automated Analysis, Technicon International Congress 1969, Vol. I, Mediad, Inc., White Plains, N.Y., 1970, pp. 333-338.

247. H. M. Etter, in Advances in Automated Analysis, Technicon International Congress 1970, Vol. II, Thurman Associates, Miami, 1971, pp. 123-126.

248. R. A. Evans and A. J. Thomas, in Instrumentation in Biochemistry (T. W. Goodwin, ed.), Academic Press, New York, 1966, pp. 69-80.

249. W. G. Evans and J. Bomstein, in Advances in Automated Analysis, Technicon International Congress 1970, Vol. II, Thurman Associates, Miami, 1971, pp. 233-239.

250. W. G. Evans, R. L. McLaughlin, and J. Bomstein, in Advances in Automated Analysis, Technicon International Congress 1969, Vol. II, Mediad, Inc., White Plains, N.Y., 1970, pp. 257-259.

251. J. W. Eveleigh, H. J. Adler, and A. S. Reichler, in Automation in Analytical Chemistry, Technicon Symposia 1967, Vol. I, Mediad, Inc., White Plains, N.Y., 1968, pp. 311-316.

252. M. A. Evenson, G. P. Hicks, J. A. Keenan, and F. C. Larson, in Automation in Analytical Chemistry, Technicon Symposia 1967, Vol. I, Mediad, Inc., White Plains, N.Y., 1968, pp. 137-140.

253. M. A. Evenson, G. P. Hicks, and R. E. Thiers, Clin. Chem., 16, 606-611 (1970).

253a. M. A. Evenson and M. A. Olson, Clin. Chem., 18, 1013-1018 (1972).

254. J. F. Failing, Jr., M. W. Buckley, and B. Zak, Amer. J. Clin. Pathol., 33, 83-88 (1960).

255. R. F. Farr, J. A. Newell, T. P. Whitehead, and G. M. Widdowson, in Automation in Analytical Chemistry, Technicon Symposia 1966, Vol. II, Mediad, Inc., White Plains, N.Y., 1967, pp. 225-229.

256. C. F. Fasce, Jr., and R. Rej, Clin. Chem., 16, 972-979 (1970).

257. S. Fassari, in Automation in Analytical Chemistry, Technicon Symposia 1967, Vol. II, Mediad, Inc., White Plains, N.Y., 1968, pp. 303-308.

258. B. Feller, W. A. Boyd, B. E. DiDario, and A. Ferrari, Ann. N.Y. Acad. Sci., 153, 428-445 (1968).

259. B. Feller, W. A. Boyd, B. E. DiDario, and A. Ferrari, in Automation in Analytical Chemistry, Technicon Symposia 1966, Vol. I, Mediad, Inc., White Plains, N.Y., 1967, pp. 206-212.

260. A. Ferrari, Ann. N.Y. Acad. Sci., 87, 792-800 (1960).

261. A. Ferrari, 1964 Technicon International Symposium, Interim Reprint, Paper No. 5, New York, 1964, "Automated Microbiological Techniques."

262. A. Ferrari, Ger. patent 1,077,895, (Cl. 42l) Mar. 17, 1960; through Chem. Abstr., 55, 17129h (1961), "Automatic Gas Analyzer for Hydrocyanic or Hydrochloric Acid Vapor."

263. A. Ferrari, U. S. patent 3,286,583 (Cl. 88-14), Nov. 22, 1966; Appl. June 22, 1962; 3 pp.; through Chem. Abstr., 66, 25879u (1967), "Colorimeter Flow Cell and Holder."

264. A. Ferrari, U. S. patent 3,074,784 (Cl. 23-253), Jan. 22, 1963; Appl. May 5, 1959; 4 pp.; through Chem. Abstr., 58, 8412f (1963), "Continuous Chromatographic Analysis."

265. A. Ferrari, Belg. patent 638,262, April 6, 1964; U. S. Appl. Oct. 18, 1962; 11 pp.; through Chem. Abstr., 63, 1501e (1965), "Continuous Solvent Extractor Design."

266. A. Ferrari, U. S. patent 3,241,923 (Cl. 23-259), March 22, 1966; Appl. Sept. 27, 1960; 6 pp.; through Chem. Abstr., 64, 16636g (1966), "Improved Apparatus and Method for Continuous Sampling and Analysis of Liquids."

267. A. Ferrari, U. S. patent 3,226,198 (Cl. 23-232), Dec. 28, 1965; Appl. Oct. 18, 1962; 6 pp.; through Chem. Abstr., 64, 10793f (1966), "Method and Apparatus for Bringing a Gas into Intimate Contact with a Thin Film of a Liquid Reagent."

268. A. Ferrari, Ger. patent 1,159,394 (Cl. B01d), Dec. 19, 1963; U. S. Appl. Nov. 17, 1959; 4 pp.; through Chem. Abstr., 60, 15460g (1964), "Ultrafilter with Cellulose Derivative Semipermeable Membrane."

269. A. Ferrari, E. Catanzaro, and F. Russo-Alesi, Ann. N.Y. Acad. Sci., 130, 602-620 (1965).

270. A. Ferrari, J. R. Gerke, R. W. Watson, and W. W. Umbreit, Ann. N.Y. Acad. Sci., 130, 704-721 (1965).

271. A. Ferrari and A. J. Khoury, Ann. N.Y. Acad. Sci., 153, 660-671 (1968).

272. A. Ferrari and G. H. MacDuff, Clin. Chem., 5, 368 (1959).

273. A. Ferrari, F. M. Russo-Alesi, and J. M. Kelly, Anal. Chem., 31, 1710-1717 (1959).

274. A. Ferrari and W. J. Smythe, U. S. patent 3,169,912 (Cl. 202-238), Feb. 16, 1965; Appl. Nov. 13, 1961; 6 pp.; through Chem. Abstr., 62, 12779e (1965), "Continuous Distillation Apparatus."

275. R. J. Ferretti and W. M. Hoffman, J. Ass. Offic. Agr. Chem., 45, 993-996 (1962).

276. B. Fingerhut, R. Ferzola, W. H. Marsh, and J. B. Levine, Ann. N.Y. Acad. Sci., 102, 137-143 (1962).

277. B. Fingerhut, A. Poock, and H. Miller, Clin. Chem., 15, 870-878 (1969).

278. V. Fiorica, Clin. Chim. Acta, 12, 191-197 (1965).

279. W. H. Fishman, J. P. Manning, M. Takeda, D. Angellis, and S. Green, Anal. Biochem., 51, 368-382 (1973).

280. R. L. Flannery and J. E. Steckel, in Soil Testing and Plant Analysis, Part I, Soil Science Society of America Special Publ. 2, Madison, Wisconsin, 1967, pp. 137-150.

280a. A. Fleck, R. D. Begg, P. I. Williams, and D. Racionzer, Ann. Clin. Biochem., 8, 13-15 (1971).

281. B. Fleet, A. Y. W. Ho, and J. Tenygl, Anal. Chem., 44, 2156-2161 (1972).

282. B. Fleet, A. Y. W. Ho, and J. Tenygl, Analyst (London), 97, 321-333 (1972).

283. B. Fleet and H. von Storp, Anal. Chem., 43, 1575-1581 (1971).

284. B. Fleet, S. Win, and T. S. West, in Automation in Analytical Chemistry, Technicon Symposia 1967, Vol. II, Mediad, Inc., White Plains, N.Y., 1968, pp. 355-359.

285. F. V. Flynn, K. A. Piper, and P. K. Roberts, J. Clin. Pathol., 19, 633-639 (1966).

286. A. Fournier, T. W. Shields, R. P. Neil, C. M. Hayes, and G. Papineau-Couture, Ann. N.Y. Acad. Sci., 153, 501-510 (1968).

287. A. Fournier, T. W. Shields, R. P. Neil, C. M. Hayes, and G. Papineau-Couture, in Automation in Analytical Chemistry, Technicon Symposia 1966, Vol. I, Mediad, Inc., White Plains, N.Y., 1967, pp. 213-217.

288. J. W. Frazer, G. D. Jones, R. Lim, M. C. Waggoner, and L. B. Rogers, Anal. Chem., 41, 1485-1487 (1969).

289. P. D. M. Frazier and G. K. Summer, Anal. Biochem., 44, 66-76 (1971).

290. H. S. Friedman, Clin. Chem., 16, 619-620 (1970).

291. S. Friedner and A. Philipson, Scand. J. Clin. Lab. Invest., 17, 185-188 (1965).

292. H. O. Friestad, D. E. Ott, and F. A. Gunther, in Advances in Automated Analysis, Technicon International Congress 1969, Vol. II, Mediad, Inc., White Plains, N.Y., 1970, pp. 21-27.

293. H. O. Friestad, D. E. Ott, and F. A. Gunther, Anal. Chem., 41, 1750-1754 (1969).

294. S. A. Fusari, D. Dittmar, and C. H. Perrizo, in Advances in Automated Analysis, Technicon International Congress 1970, Vol. II, Thurman Associates, Miami, 1971, pp. 241-245.

295. R. H. Gaddy, Jr., in Automation in Analytical Chemistry, Technicon Symposia 1965, Mediad, Inc., New York, 1966, pp. 210-214.

296. R. H. Gaddy, Jr., in Automation in Analytical Chemistry, Technicon Symposia 1965, Mediad, Inc., New York, 1966, pp. 215-219.

297. R. H. Gaddy and R. S. Dorsett, Anal. Chem., 40, 429-432 (1968).

298. P. Gale, P. A. I. George, and W. D. Williams, in Automation in Analytical Chemistry, Technicon Symposia 1967, Vol. II, Mediad, Inc., White Plains, N.Y., 1968, pp. 411-415.

299. M. E. Gales, Jr., W. H. Kaylor, and J. E. Longbottom, Analyst (London), 93, 97-100 (1968).

300. M. R. Galley, in Automation in Analytical Chemistry, Technicon Symposia 1966, Vol. I, Mediad, Inc., White Plains, N.Y., 1967, pp. 643-648.

301. S. R. Gambino, Amer. J. Clin. Pathol., 35, 268-269 (1961).

302. S. R. Gambino and H. Schreiber, Amer. J. Clin. Pathol., 45, 406-411 (1966).

303. F. M. Ganis, in Advances in Automated Analysis, Technicon International Congress 1970, Vol. I, Thurman Associates, Miami, 1971, pp. 551-553.

304. F. M. Ganis, C. A. Berger, C. J. Migeon, and T. B. Connor, in Automation in Analytical Chemistry, Technicon Symposia 1966, Vol. I, Mediad, Inc., White Plains, N.Y., 1967, pp. 83-86.

305. W. M. Gantenbein, J. Ass. Offic. Anal. Chem., 55, 123-127 (1972).

306. S. A. Gardanier and G. H. Spooner, Amer. J. Clin. Pathol., 54, 341-347 (1970).

307. E. S. Garnett, A. C. Pollard, and C. E. Webber, Lancet, 2, 318-319 (1965).

308. P. G. Garritsen, E. G. Beals, and B. Meddings, in Automation in Analytical Chemistry, Technicon Symposia 1966, Vol. I, Mediad, Inc., White Plains, N.Y., 1967, pp. 604-606.

309. P. I. Garry and G. M. Owen, in Automation in Analytical Chemistry, Technicon Symposia 1967, Vol. I, Mediad, Inc., White Plains, N.Y., 1968, pp. 507-510.

310. A. C. Garza and H. E. Weissler, in Automation in Analytical Chemistry, Technicon Symposia 1967, Vol. I, Mediad, Inc., White Plains, N.Y., 1968, pp. 201-206.

311. M. A. Gauger and L. M. White, Anal. Biochem., 35, 130-135 (1970).

312. A. P. Gaunce and A. D'Iorio, Anal. Biochem., 37, 204-207 (1970).

313. J. M. Gawthorne, J. Watson, and E. L. R. Stokstad, Anal. Biochem., 42, 555-559 (1971).

314. C. W. Gehrke, J. H. Baumgartner, and J. P. Ussary, J. Ass. Offic. Anal. Chem., 49, 1213-1218 (1966).

315. C. W. Gehrke, F. E. Kaiser, and J. P. Ussary, in Automation in Analytical Chemistry, Technicon Symposia 1967, Vol. I, Mediad, Inc., White Plains, N.Y., 1968, pp. 239-251.

316. C. W. Gehrke, F. E. Kaiser, and J. P. Ussary, J. Ass. Offic. Anal. Chem., 51, 200-211 (1968).

317. C. W. Gehrke, J. S. Killingley, and L. L. Wall, Sr., J. Ass. Offic. Anal. Chem., 55, 467-480 (1972).

318. C. W. Gehrke, J. P. Ussary, and J. H. Baumgartner, in Automation in Analytical Chemistry, Technicon Symposia 1966, Vol. I, Mediad, Inc., White Plains, N.Y., 1967, pp. 171-176.

319. C. W. Gehrke, J. P. Ussary, and G. H. Kramer, Jr., J. Ass. Offic. Agr. Chem., 47, 459-469 (1964).

320. C. W. Gehrke, L. L. Wall, Sr., J. S. Killingley, and K. Inada, in Advances in Automated Analysis, Technicon International Congress 1970, Vol. II, Thurman Associates, Miami, 1971, pp. 39-45.

321. C. W. Gehrke, L. L. Wall, Sr., J. S. Killingley, and K. Inada, J. Ass. Offic. Anal. Chem., 54, 651-657 (1971).

322. D. L. Geiger and E. H. Vernot, in Automation in Analytical Chemistry, Technicon Symposia 1967, Vol. I, Mediad, Inc., White Plains, N.Y., 1968, pp. 427-430.

323. J. R. Gerke and A. Ferrari, in Automation in Analytical Chemistry, Technicon Symposia 1967, Vol. I, Mediad, Inc., White Plains, N.Y., 1968, pp. 531-540.

324. J. R. Gerke, T. A. Haney, J. F. Pagano, and A. Ferrari, Ann. N.Y. Acad. Sci., 87, 782-791 (1960).

325. E. M. Gindler, Clin. Chem., 16, 350-351 (1970).

326. T. J. Giovanniello, G. DiBenedetto, D. W. Palmer, and T. Peters, Jr., in Automation in Analytical Chemistry, Technicon Symposia 1967, Vol. I, Mediad, Inc., White Plains, N.Y., 1968, pp. 185-188.

327. T. J. Giovanniello, G. DiBenedetto, D. W. Palmer, and T. Peters, Jr., J. Lab. Clin. Med., 71, 874-883 (1968).

328. H. J. Gitelman, Anal. Biochem., 18, 521-531 (1967).

329. H. J. Gitelman, C. Hurt, and L. Lutwak, Anal. Biochem., 14, 106-120 (1966).

330. H. J. Gitelman, C. Hurt, and L. Lutwak, in Automation in Analytical Chemistry, Technicon Symposia 1965, Mediad, Inc., New York, 1966, pp. 352-355.

331. N. Gochman, in Automation in Analytical Chemistry, Technicon Symposia 1965, Mediad, Inc., New York, 1966, pp. 528-531.

332. N. Gochman, 1964 Technicon International Symposium, Interim Reprint, New York, "Operation of the Technicon Flame Photometer with Low Pressure Natural Gas."

333. N. Gochman and H. Givelber, Clin. Chem., 16, 229-234 (1970).

334. G. Goldstein, W. L. Maddox, and I. B. Rubin, in Automation in Analytical Chemistry, Technicon Symposia 1967, Vol. I, Mediad, Inc., White Plains, N.Y., 1968, pp. 47-49.

335. R. R. Goodall and R. Davies, Analyst (London), 86, 326-335 (1961).

336. H. E. Gould, in Automation in Analytical Chemistry, Technicon Symposia 1967, Vol. I, Mediad, Inc., White Plains, N.Y., 1968, pp. 141-146.

337. H. E. Gould and A. Brooks, in Automation in Analytical Chemistry, Technicon Symposia 1966, Vol. I, Mediad, Inc., White Plains, N.Y., 1967, pp. 378-382.

338. P. D. Goulden and B. K. Afghan, in Advances in Automated Analysis, Technicon International Congress 1970, Vol. II, Thurman Associates, Miami, 1971, pp. 317-321.

339. P. D. Goulden, B. K. Afghan, and P. Brooksbank, Anal. Chem., 44, 1845-1849 (1972).

340. H. J. Grady and M. A. Lamar, Clin. Chem., 5, 542-550 (1959).

341. R. Grasbeck and R. Karlsson, Acta Chem. Scand., 17, 1-7 (1963).

342. K. Grasshoff, in Advances in Automated Analysis, Technicon International Congress 1969, Vol. II, Mediad, Inc., White Plains, N.Y., 1970, pp. 133-145.

343. K. Grasshoff, in Automation in Analytical Chemistry, Technicon Symposia 1965, Mediad, Inc., New York, 1966, pp. 304-307.

344. K. Grasshoff, in Automation in Analytical Chemistry, Technicon Symposia 1966, Vol. I, Mediad, Inc., White Plains, N.Y., 1967, pp. 573-579.

345. P. Gray and J. A. Owen, Clin. Chim. Acta, 24, 389-399 (1969).

346. V. J. Greely, W. W. Holl, T. P. Michaels, and L. P. Sinotte, Ann. N.Y. Acad. Sci., 130, 657-663 (1965).

347. R. M. Greenway, in Automation in Analytical Chemistry, Technicon Symposia 1965, Mediad, Inc., New York, 1966, pp. 463-465.

348. J. D. Gregory and L. Van Lenten, in Automation in Analytical Chemistry, Technicon Symposia 1965, Mediad, Inc., New York, 1966, pp. 620-625.

349. P. D. Griffiths and N. W. Carter, J. Clin. Pathol., 22, 609-616 (1969).

350. A. Grossman, G. F. Grossman, R. L. Pollack, and E. Kravitz, Anal. Biochem., 8, 124-126 (1964).

351. P. W. Grunmeier, A. Gray, and A. Ferrari, Ann. N.Y. Acad. Sci., 130, 809-818 (1965).

352. R. Guillaumot, in Automation in Analytical Chemistry, Technicon Symposia 1965, Mediad, Inc., New York, 1966, pp. 396-398.

353. F. A. Gunther and D. E. Ott, Analyst (London), 91, 475-481 (1966).

354. F. A. Gunther and D. E. Ott, in Automation in Analytical Chemistry, Technicon Symposia 1965, Mediad, Inc., New York, 1966, pp. 96-108.

355. F. A. Gunther and D. E. Ott, in Residue Reviews (F. A. Gunther, ed.), Vol. 14, Springer-Verlag, New York, 1966, pp. 12-38.

356. R. L. Habig, B. W. Schelin, L. Walters, and R. E. Thiers, in Advances in Automated Analysis, Technicon International Congress 1969, Vol. I, Mediad, Inc., White Plains, N.Y., 1970, pp. 139-143.

357. R. L. Habig, B. W. Schlein, L. Walters, and R. E. Thiers, Clin. Chem., 15, 1045-1055 (1969).

358. R. L. Habig and W. R. Williamson, Clin. Chem., 16, 251-253 (1970).

359. F. C. Hadley, in Automation in Analytical Chemistry, Technicon Symposia 1965, Mediad, Inc., New York, 1966, pp. 383-386.

360. J. P. Haesen, G. T. Berends, and H. A. Zondag, Clin. Chim. Acta, 37, 463-470 (1972).

361. A. C. Haff, Clin. Chem., 17, 1134-1136 (1971).

362. G. D. Haines and C. D. Anselm, in Advances in Automated Analysis, Technicon International Congress 1969, Vol. II, Mediad, Inc., White Plains, N.Y., 1970, pp. 159-162.

363. A. Hainline, Jr., Cleveland Clin. Quart., 30, 131-145 (1963).

364. I. R. Hainsworth and P. E. Hall, Clin. Chim. Acta, 35, 201-208 (1971).

365. J. R. S. Hales, A. Little, and M. E. D. Webster, Anal. Biochem., 16, 114-118 (1966).

366. C. H. Hall and R. C. Meyer, in Advances in Automated Analysis, Technicon International Congress 1970, Vol. II, Thurman Associates, Miami, 1971, pp. 215-218.

367. J. W. Hall and D. M. Tucker, Anal. Biochem., 26, 12-17 (1968).

368. R. A. Hall and T. P. Whitehead, J. Clin. Pathol., 23, 323-326 (1970).

369. L. G. Hambleton, in Advances in Automated Analysis, Technicon International Congress 1970, Vol. II, Thurman Associates, Miami, 1971, pp. 53-56.

370. L. G. Hambleton, J. Ass. Offic. Anal. Chem., 53, 456-460 (1970).

371. L. G. Hambleton, J. Ass. Offic. Anal. Chem., 54, 646-650 (1971).

372. P. B. Hamilton, Ann. N.Y. Acad. Sci., 102, 55-75 (1962).

373. R. B. Hanawalt and J. E. Steckel, in Automation in Analytical Chemistry, Technicon Symposia 1966, Vol. I, Mediad, Inc., White Plains, N.Y., 1967, pp. 133-136.

374. T. A. Haney, J. R. Gerke, M. E. Madigan, J. F. Pagano, and A. Ferrari, Ann. N.Y. Acad. Sci., 93, 627-639 (1962).

375. A. Hanok and J. Kuo, in Automation in Analytical Chemistry, Technicon Symposia 1966, Vol. I, Mediad, Inc., White Plains, N.Y., 1967, pp. 336-344.

376. R. J. Hansen, A. R. Tschida, and H. Markowitz, Clin. Chem., 18, 677-678 (1972).

377. M. R. Hardeman, A. Den Uyl, and H. K. Prins, Clin. Chim. Acta, 37, 71-79 (1972).

378. J. L. Hardman, in Advances in Automated Analysis, Technicon International Congress 1970, Vol. II, Thurman Associates, Miami, 1971, pp. 323-332.

379. P. W. Hathaway, L. Jakoi, W. G. Troyer, Jr., and M. D. Bogdonoff, Anal. Biochem., 20, 466-476 (1967).

380. G. G. Hazen, J. A. Hause, and J. A. Hubicki, Ann. N.Y. Acad. Sci., 130, 761-768 (1965).

381. W. Heerspink and G. J. Op de Weegh, Clin. Chim. Acta, 29, 191-192 (1970).

382. R. M. Heinicke, C. Larson, O. Levand, and M. McCarter, in Automation in Analytical Chemistry, Technicon Symposia 1967, Vol. I, Mediad, Inc., White Plains, N.Y., 1968, pp. 207-212.

383. G. W. Heinke and H. Behmann, in Advances in Automated Analysis, Technicon International Congress 1969, Vol. II, Mediad, Inc., White Plains, N.Y., 1970, pp. 115-120.

384. R. N. Heistand, Anal. Chem., 42, 903-906 (1970).

385. S. Hellerstein, H. Grady, and S. Grisolia, J. Lab. Clin. Med., 76, 171-174 (1970).

386. W. Helmreich, in Advances in Automated Analysis, Technicon International Congress 1969, Vol. I, Mediad, Inc., White Plains, N.Y., 1970, pp. 39-41.

387. G. Henderson and A. Gajjar, J. Chem. Educ., 48, 693-694 (1971).

388. A. Henriksen, Analyst (London), 90, 29-34 (1965).

389. A. Henriksen, in Automation in Analytical Chemistry, Technicon Symposia 1965, Mediad, Inc., New York, 1966, pp. 301-303.

390. A. Henriksen and A. R. Selmer-Olsen, Analyst (London), 95, 514-518 (1970).

391. G. Herman, III, E. L. Cohen, and H. T. Sugiura, Amer. J. Clin. Pathol., 54, 226-230 (1970).

392. R. F. Heuermann, R. C. Kroner, V. C. Midkiff, G. Schwartzman, and H. P. Eiduson, J. Ass. Offic. Anal. Chem., 55, 368-369 (1972).

393. G. P. Hicks, M. M. Gieschen, and D. J. Nichols, Clin. Chem., 15, 811-812 (1969).

394. J. B. Hill, Clin. Chem., 11, 122-130 (1965).

395. J. B. Hill and D. S. Cowart, Anal. Biochem., 16, 327-337 (1966).

396. J. B. Hill and G. Kessler, J. Lab. Clin. Med., 57, 970-980 (1961).

397. R. L. Hill and R. Delaney, in Methods in Enzymology (C. H. W. Hirs, ed.), Vol. 11, Academic Press, New York, 1967, pp. 339-351.

398. A. Himoe, G. L. Catledge, and W. E. Kurtin, Anal. Chem., 44, 391-394 (1972).

399. R. H. Hinton and K. A. Norris, Anal. Biochem., 48, 247-258 (1972).

400. N. J. Hochella, Anal. Biochem., 21, 227-234 (1967).

401. N. J. Hochella, in Automation in Analytical Chemistry, Technicon Symposia 1966, Vol. I, Mediad, Inc., White Plains, N.Y., 1967, pp. 29-32.

402. N. J. Hochella and J. B. Hill, in Advances in Automated Analysis, Technicon International Congress 1969, Vol. I, Mediad, Inc., White Plains, N.Y., 1970, pp. 55-60.

403. N. J. Hochella and J. B. Hill, Clin. Chem., 15, 949-960 (1969).

404. N. J. Hochella and S. Weinhouse, Anal. Biochem., 10, 304-317 (1965).

405. N. J. Hochella and S. Weinhouse, in Automation in Analytical Chemistry, Technicon Symposia 1965, Mediad, Inc., New York, 1966, pp. 539-544.

406. M. Hoffman, A. Bar-Akiva, and L. Tanhum, Anal. Biochem., 38, 35-39 (1970).

407. W. W. Holl, J. H. Tufekjian, T. P. Michaels, and L. P. Sinotte, Ann. N.Y. Acad. Sci., 130, 525-531 (1965).

408. W. W. Holl and R. W. Walton, Ann. N.Y. Acad. Sci., 130, 504-515 (1965).

409. W. R. Holub and F. A. Galli, Clin. Chem., 18, 239-243 (1972).

410. H. W. Holy, 1963 Technicon International Symposium, London, pp. 9-14.

411. J. K. Hoober and I. A. Bernstein, Anal. Biochem., 9, 467-473 (1964).

412. D. A. Hopkinson and W. H. P. Lewis, in Automation in Analytical Chemistry, Technicon Symposia 1967, Vol. II, Mediad, Inc., White Plains, N.Y., 1968, pp. 227-231.

413. W. D. Hormann, G. Formica, K. Ramsteiner, and D. O. Eberle, J. Ass. Offic. Anal. Chem., 55, 1031-1038 (1972).

414. D. B. Horn, Clin. Chim. Acta, 37, 43-46 (1972).

415. H. F. Hosley, K. B. Olson, J. Horton, P. Michelsen, and R. Atkins, in Advances in Automated Analysis, Technicon International Congress 1969, Vol. I, Mediad, Inc., White Plains, N.Y., 1970, pp. 105-110.

416. J. L. Hoyt and D. E. Jordan, in Advances in Automated Analysis, Technicon International Congress 1969, Vol. II, Mediad, Inc., White Plains, N.Y., 1970, pp. 73-77.

417. J. L. Hoyt and D. E. Jordan, J. Ass. Offic. Anal. Chem., 52, 1121-1126 (1969).

418. R. P. Huemer and K.-D. Lee, Anal. Biochem., 37, 149-153 (1970).

419. J. A. Hunt, Anal. Biochem., 23, 289-300 (1968).

420. D. T. Hunter, L. D. McGuire, and V. Degn, Amer. J. Clin. Pathol., 50, 723-725 (1968).

421. R. J. Hurst and A. M. Bold, J. Clin. Pathol., 19, 622-624 (1966).

422. R. L. Hussey, J. L. Hale, and D. P. Howard, J. Pharm. Sci., 62, 1171-1173 (1973).

423. K. Hviid, Clin. Chem., 13, 281-289 (1967).

424. G. B. P. Ingram, J. Clin. Pathol., 23, 187-188 (1970).

425. W. T. Ingram, J. Golden, E. J. Kaplin, M. P. Levine, and R. R. Cardenas, Jr., in Automation in Analytical Chemistry, Technicon Symposia 1967, Vol. I, Mediad, Inc., White Plains, N.Y., 1968, pp. 409-415.

426. W. J. Irvine, in Automation in Analytical Chemistry, Technicon Symposia 1966, Vol. II, Mediad, Inc., White Plains, N.Y., 1967, pp. 347-353.

427. W. J. Irvine and K. J. G. Marwick, in Automation in Analytical Chemistry, Technicon Symposia 1967, Vol. II, Mediad, Inc., White Plains, N.Y., 1968, pp. 33-37.

428. J. Isreeli, in Advances in Automated Analysis, Technicon International Congress 1970, Vol. I, Thurman Associates, Miami, 1971, pp. 81-83.

429. J. Isreeli, M. Pelavin, and G. Kessler, Ann. N.Y. Acad. Sci., 87, 636-649 (1960).

430. H. Jacobson, Anal. Chem., 38, 1951-1954 (1966).

431. H. Jacobson, Ann. N.Y. Acad. Sci., 153, 486-492 (1968).

432. V. H. T. James and J. Townsend, in Automation in Analytical Chemistry, Technicon Symposia 1967, Vol. I, Mediad, Inc., White Plains, N.Y., 1968, pp. 41-45.

433. A. M. Jamieson, A. Buday, and G. VanGheluwe, in Automation in Analytical Chemistry, Technicon Symposia 1965, Mediad, Inc., New York, 1966, pp. 76-77.

434. A. M. Jamieson and C. G. MacKinnon, in Automation in Analytical Chemistry, Technicon Symposia 1965, Mediad, Inc., New York, 1966, pp. 73-75.

435. A. P. Jansen, K. A. Peters, and T. Zelders, Clin. Chim. Acta, 27, 125-132 (1970).

436. M. Jellinek, H. Amako, and V. Willman, in Advances in Automated Analysis, Technicon International Congress 1970, Vol. I, Thurman Associates, Miami, 1971, pp. 587-590.

437. H. Jenner, in Automation in Analytical Chemistry, Technicon Symposia 1967, Vol. II, Mediad, Inc., White Plains, N.Y., 1968, pp. 203-207.

438. P. C. Jocelyn, Biochem. J., 85, 480-485 (1962).

439. E. A. Johnson, D. A. Rigas, and R. T. Jones, in Automation in Analytical Chemistry, Technicon Symposia 1966, Vol. I, Mediad, Inc., White Plains, N.Y., 1967, pp. 410-415.

440. A. Jones and G. Palmer, Analyst (London), 95, 463-465 (1970).

441. R. T. Jones, in Automation in Analytical Chemistry, Technicon Symposia 1966, Vol. I, Mediad, Inc., White Plains, N.Y., 1967, pp. 416-423.

442. R. T. Jones, in Methods of Biochemical Analysis (D. Glick, ed.), Vol. 18, Wiley (Interscience), New York, 1970, pp. 205-258.

443. D. E. Jordan, in Automation in Analytical Chemistry, Technicon Symposia 1967, Vol. I, Mediad, Inc., White Plains, N.Y., 1968, pp. 253-256.

444. D. E. Jordan, J. Ass. Offic. Anal. Chem., 52, 581-587 (1969).

445. B. N. Kabadi, in Advances in Automated Analysis, Technicon International Congress 1970, Vol. II, Thurman Associates, Miami, 1971, pp. 247-252.

446. B. N. Kabadi, J. Pharm. Sci., 60, 1862-1865 (1971).

447. B. N. Kabadi, A. T. Warren, and C. H. Newman, in Advances in Automated Analysis, Technicon International Congress 1969, Vol. II, Mediad, Inc., White Plains, N.Y., 1970, pp. 207-212.

448. B. N. Kabadi, A. T. Warren, and C. H. Newman, J. Pharm. Sci., 58, 1127-1131 (1969).

449. A. H. Kadish, Amer. J. Med. Electron., 3, 82-86 (1964).

450. A. H. Kadish and D. A. Hall, Clin. Chem., 11, 869-875 (1965).

451. P. P. Kamoun, J. M. Pleau, and N. K. Man, Clin. Chem., 18, 355-357 (1972).

452. A. E. Kaptionak, E. Biernacka, and H. J. Pazdera, in Automation in Analytical Chemistry, Technicon Symposia 1965, Mediad, Inc., New York, 1966, pp. 27-33.

453. S. Kashket, in Advances in Automated Analysis, Technicon International Congress 1969, Vol. I, Mediad, Inc., White Plains, N.Y., 1970, pp. 17-19.

454. G. M. Katz and A. L. Levy, Clin. Chem., 18, 596 (1972).

455. V. Kauppinen and C-G. Gref, Scand. J. Clin. Lab. Invest., 20, 24-28 (1967).

456. E. Kawerau, in Advances in Automated Analysis, Technicon International Congress 1969, Vol. I, Mediad, Inc., White Plains, N.Y., 1970, pp. 165-169.

457. E. Kawerau, in Automation in Analytical Chemistry, Technicon Symposia 1966, Vol. II, Mediad, Inc., White Plains, N.Y., 1967, pp. 293-301.

458. J. Keay, Analyst (London), 94, 690-694 (1969).

459. J. Keay and P. M. A. Menage, Analyst (London), 94, 895-899 (1969).

460. J. Keay, P. M. A. Menage, and G. A. Dean, Analyst (London), 97, 897-902 (1972).

461. P. W. Kelley and F. D. Fuller, in Automation in Analytical Chemistry, Technicon Symposia 1965, Mediad, Inc., New York, 1966, pp. 266-269.

462. J. M. Kelly, IRE Trans. Bio-Med. Electron., 8, 98-108 (1961).

463. J. R. Kelly and L. C. Wilson, Ann. N.Y. Acad. Sci., 130, 575-581 (1965).

464. J. H. Kemp, J. Clin. Pathol., 19, 400-401 (1966).

465. A. P. Kendal, Anal. Biochem., 23, 150-155 (1968).

466. A. P. Kendal, in Automation in Analytical Chemistry, Technicon Symposia 1967, Vol. II, Mediad, Inc., White Plains, N.Y., 1968, pp. 175-181.

467. A. P. Kenny and A. Jamieson, Clin. Chim. Acta, 10, 536-543 (1964).

468. M. A. Kenny and M. H. Cheng, Clin. Chem., 18, 352-354 (1972).

469. M. Kenny, P. A. Van Dreal, and A. Kaplan, Clin. Chem., 15, 763-764 (1969).

470. R. B. Kesler, Anal. Chem., 35, 963-965 (1963).

471. R. B. Kesler, Anal. Chem., 39, 1416-1422 (1967).

472. R. B. Kesler, in Automation in Analytical Chemistry, Technicon Symposia 1965, Mediad, Inc., New York, 1966, pp. 174-177.

473. G. Kessler, in Advances in Clinical Chemistry (O. Bodansky and C. P. Stewart, eds.), Vol. 10, Academic Press, New York, 1967, pp. 45-64.

474. G. Kessler, in Gradwohl's Clinical Laboratory Methods and Diagnosis, 6th ed. (S. Frankel, S. Reitman, and A. C. Sonnenwirth, eds.), Vol. I, C. V. Mosby Co., St. Louis, 1963, pp. 301-313.

475. G. Kessler, in Gradwohl's Clinical Laboratory Methods and Diagnosis, 7th ed. (S. Frankel, S. Reitman, and A. C. Sonnenwirth, eds.), Vol. I, C. V. Mosby Co., St. Louis, 1970, pp. 323-372.

476. G. Kessler, R. L. Rush, L. Leon, A. Delea, and R. Cupiola, in Advances in Automated Analysis, Technicon International Congress 1970, Vol. I, Thurman Associates, Miami, 1971, pp. 67-74.

477. G. Kessler and M. Wolfman, Clin. Chem., 10, 686-703 (1964).

478. A. J. Khoury, in Automation in Analytical Chemistry, Technicon Symposia 1966, Vol. I, Mediad, Inc., White Plains, N.Y., 1967, pp. 192-195.

479. A. J. Khoury and L. J. Cali, Ann. N.Y. Acad. Sci., 153, 456-460 (1968).

480. F. C. A. Killer, in Automation in Analytical Chemistry, Technicon Symposia 1966, Vol. I, Mediad, Inc., White Plains, N.Y., 1967, pp. 652-655.

481. L. M. Killingsworth and J. Savory, Clin. Chem., 17, 936-940 (1971).

482. L. M. Killingsworth, J. Savory, and P. O. Teague, Clin. Chem., 17, 374-377 (1971).

483. J. C. Kirschman, J. J. Iacono, and W. G. Jones, in Automation in Analytical Chemistry, Technicon Symposia 1967, Vol. I, Mediad, Inc., White Plains, N.Y., 1968, pp. 519-525.

484. B. Klein, J. Auerbach, and S. Morgenstern, Clin. Chem., 11, 998-1008 (1965).

485. B. Klein and J. H. Kaufman, Clin. Chem., 13, 290-298 (1967).

486. B. Klein and J. H. Kaufman, Clin. Chem., 13, 1079-1087 (1967).

487. B. Klein, J. H. Kaufman, and S. Morgenstern, in Automation in Analytical Chemistry, Technicon Symposia 1966, Vol. I, Mediad, Inc., White Plains, N.Y., 1967, pp. 10-14.

488. B. Klein, J. H. Kaufman, and S. Morgenstern, Clin. Chem., 13, 388-396 (1967).

489. B. Klein, J. H. Kaufman, and M. Oklander, Clin. Chem., 13, 788-796 (1967).

490. B. Klein, J. H. Kaufman, and M. Oklander, Clin. Chem., 13, 797-805 (1967).

491. B. Klein and M. Oklander, Clin. Chem., 13, 26-35 (1967).

492. M. H. Kline, F. Medzihradsky, and L. E. Hokin, Anal. Biochem.,
      23, 97-101 (1968).

493. C. F. Knowles and A. Hodgkinson, Analyst (London), 97, 474-481
      (1972).

494. H. Ko and M. E. Royer, Anal. Biochem., 26, 18-33 (1968).

495. J. Korf, H. H. Schutte, and K. Venema, Anal. Biochem., 53,
      146-153 (1973).

496. A. Koszyn, E. Heier, and M. Pelavin, in Advances in Automated
      Analysis, Technicon International Congress 1970, Vol. I, Thurman
      Associates, Miami, 1971, pp. 57-62.

497. D. G. Kramme, R. H. Griffen, C. G. Hartford, and J. A. Corrado,
      Anal. Chem., 45, 405-408 (1973).

498. J. Kream, B. Thysen, A. Griogorian, and L. Hellman, in Advances
      in Automated Analysis, Technicon International Congress 1970,
      Vol. I, Thurman Associates, Miami, 1971, pp. 555-561.

499. M. I. Krichevsky, J. Schwartz, and M. Mage, Anal. Biochem., 12,
      94-105 (1965).

500. A. F. Krieg and B. A. Hutchinson, Clin. Chem., 16, 443-445 (1970).

501. H. Kruijswijk and J. G. Pelle, Clin. Chim. Acta, 8, 321-323 (1963).

502. C. H. Kurtzman, P. Smith, Jr., and D. G. Snyder, Anal. Biochem.,
      12, 282-289 (1965).

503. E. J. Kusner and D. J. Herzig, in Advances in Automated Analysis,
      Technicon International Congress 1970, Vol. II, Thurman Associates,
      Miami, 1971, pp. 429-432.

504. H. Kutzim, in Automation in Analytical Chemistry, Technicon
      Symposia 1965, Mediad, Inc., New York, 1966, pp. 393-395.

505. N. Kuzel, J. Pharm. Sci., 57, 852-855 (1968).

506. N. R. Kuzel, Ann. N.Y. Acad. Sci., 130, 858-868 (1965).

507. N. R. Kuzel, in Automation in Analytical Chemistry, Technicon
      Symposia 1966, Vol. I, Mediad, Inc., White Plains, N.Y., 1967,
      pp. 218-221.

508. N. R. Kuzel and H. F. Coffey, in Automation in Analytical
      Chemistry, Technicon Symposia 1966, Vol. I, Mediad, Inc., White
      Plains, N.Y., 1967, pp. 235-239.

509. N. R. Kuzel and H. F. Coffey, J. Pharm. Sci., 56, 522-525 (1967).

510. N. R. Kuzel and H. E. Roudebush, Ann. N.Y. Acad. Sci., 153, 416-427 (1968).

511. N. R. Kuzel, H. E. Roudebush, and C. E. Stevenson, J. Pharm. Sci., 58, 381-406 (1969).

512. J. Lacy, Analyst (London), 90, 65-75 (1965).

513. W. W. Lacy and O. B. Crofford, J. Lab. Clin. Med., 64, 828-836 (1964).

514. R. H. Laessig, T. H. Schwartz, and T. A. Paskey, Clin. Chem., 18, 48-51 (1972).

515. R. H. Laessig, P. P. Tong, and G. G. Hoffman, Clin. Chem., 15, 813 (1969).

516. R. H. Laessig, C. E. Underwood, and B. J. Basteyns, Clin. Chem., 13, 985-993 (1967).

517. E. A. Lane and C. Mavrides, Anal. Biochem., 27, 363-366 (1969).

518. J. R. Lane, J. Ass. Offic. Anal. Chem., 54, 596-599 (1971).

519. B. M. Lapidus, J. W. Tetrud, and A. Karmen, Clin. Chem., 17, 231-232 (1971).

520. C. Larson, S. J. Dowden, and J. M. Gorman, in Advances in Automated Analysis, Technicon International Congress 1970, Vol. I, Thurman Associates, Miami, 1971, pp. 347-353.

521. L.-I. Larsson and O. Samuelson, Acta Chem. Scand., 19, 1357-1364 (1965).

522. L.-I. Larsson and O. Samuelson, Mikrochim. Acta, 1967, 328-332 (1967).

523. R. Lauwerys, R. Delbroeck, and M. D. Vens, Clin. Chim. Acta, 40, 443-447 (1972).

524. A. R. Law, N. J. Nicolson, and R. L. Norton, J. Ass. Offic. Anal. Chem., 54, 764-768 (1971).

525. A. L. Lazrus, K. C. Hill, and J. P. Lodge, in Automation in Analytical Chemistry, Technicon Symposia 1965, Mediad, Inc., New York, 1966, pp. 291-293.

526. Y. C. Lee, J. F. McKelvy, and D. Lang, Anal. Biochem., 27, 567-574 (1969).

527. Y. C. Lee, J. R. Scocca, and L. Muir, Anal. Biochem., 27, 559-566 (1969).

528. J. Lenard, S. L. Johnson, R. W. Hyman, and G. P. Hess, Anal. Biochem., 11, 30-41 (1965).

529. H. G. Lento, in Automation in Analytical Chemistry, Technicon Symposia 1966, Vol. I, Mediad, Inc., White Plains, N.Y., 1967, pp. 598-603.

530. H. G. Lento and C. E. Daugherty, in Advances in Automated Analysis, Technicon International Congress 1970, Vol. II, Thurman Associates, Miami, 1971, pp. 75-80.

531. J. Levine, 1964 Technicon International Symposium, Interim Reprint, New York, "Multiple Channel Approach to Mass Screening."

532. J. B. Levine and E. W. Larrabee, in Automation in Analytical Chemistry, Technicon Symposia 1966, Vol. I, Mediad, Inc., White Plains, N.Y., 1967, pp. 15-17.

533. J. B. Levine and B. Zak, Clin. Chim. Acta, 10, 381-384 (1964).

534. M. Levine, F. E. Dorer, J. R. Kahn, K. E. Lentz, and L. T. Skeggs, Anal. Biochem., 34, 366-375 (1970).

535. A. L. Levy and D. Kanon, Clin. Chem., 15, 813-814 (1969).

536. A. L. Levy and C. Keyloun, in Advances in Automated Analysis, Technicon International Congress 1970, Vol. I, Thurman Associates, Miami, 1971, pp. 497-502.

537. A. L. Levy and S. Konig-Levy, Clin. Chem., 18, 1539-1540 (1972).

538. H. Lindley and T. Haylett, J. Chromatogr., 32, 193-194 (1968).

539. J. E. Lindquist, Anal. Chim. Acta, 41, 158-160 (1968).

540. J. G. Lines, J. Clin. Pathol., 22, 617-620 (1969).

541. R. B. Lingeman and A. W. Musser, in Standard Methods of Chemical Analysis, 6th ed. (F. J. Welcher, ed.), Part B, Vol. 3, Van Nostrand, Princeton, N.J., 1966, pp. 975-1019.

542. E. W. Linton, L. C. Rodgers, and R. G. Garcia, in Automation in Analytical Chemistry, Technicon Symposia 1967, Vol. I, Mediad, Inc., White Plains, N.Y., 1968, pp. 487-490.

543. L. J. Lionnel, Analyst (London), 95, 194-199 (1970).

544. M. H. Litchfield, Analyst (London), 92, 132-136 (1967).

545. R. A. Llenado and G. A. Rechnitz, Anal. Chem., 45, 826-833 (1973).

546. E. E. Logsdon, Ann. N.Y. Acad. Sci., 87, 801-807 (1960).

547. A. Looye, Clin. Chem., 16, 753-755 (1970).

548. A. Looye, E. W. Kwarts, and A. Groen, Clin. Chem., 14, 890-897 (1968).

549. A. Lopez, R. Vial, L. Gremillion, and L. Bell, Clin. Chem., 17, 994-997 (1971).

550. E. Lorch, Anal. Biochem., 28, 307-312 (1969).

551. E. Lorch and K. F. Gey, Anal. Biochem., 16, 244-252 (1966).

552. J. A. Lott and T. S. Herman, Clin. Chem., 17, 614-621 (1971).

553. R. K. Love and M. E. McCoy, in Advances in Automated Analysis, Technicon International Congress 1969, Vol. II, Mediad, Inc., White Plains, N.Y., 1970, pp. 239-243.

554. D. P. Lundgren, Ann. N.Y. Acad. Sci., 87, 904-910 (1960).

555. D. P. Lundgren and N. P. Loeb, Anal. Chem., 33, 366-370 (1961).

556. C. C. Mabry, R. E. Gevedon, I. E. Roeckel, and N. Gochman, Amer. J. Clin. Pathol., 46, 265-281 (1966).

557. C. C. Mabry, R. E. Gevedon, I. E. Roeckel, and N. Gochman, in Automation in Analytical Chemistry, Technicon Symposia 1966, Vol. I, Mediad, Inc., White Plains, N.Y., 1967, pp. 18-28.

558. C. C. Mabry, R. E. Gevedon, I. E. Roeckel, and N. Gochman, Tech. Bull. Regist. Med. Technol. Amer. Soc. Clin. Pathol., 36, 161-177 (1966).

559. R. R. Macari and J. F. Williams, in Advances in Automated Analysis, Technicon International Congress 1969, Vol. II, Mediad, Inc., White Plains, N.Y., 1970, pp. 229-231.

560. A. MacDonald, M. Dawson, E. Castro, R. Cunningham, J. Westheimer, and M. J. Walsh, in Advances in Automated Analysis, Technicon International Congress 1970, Vol. II, Thurman Associates, Miami, 1971, pp. 219-226.

561. E. A. MacLean, Amer. J. Clin. Pathol., 51, 427-428 (1969).

562. R. H. Mandl, in Protides of the Biological Fluids (H. Peeters, ed.), Elsevier, Amsterdam, 1962, pp. 81-85.

563. R. H. Mandl, J. F. Goldman, and L. H. Weinstein, in Automation in Analytical Chemistry, Technicon Symposia 1966, Vol. I, Mediad, Inc., White Plains, N.Y., 1967, pp. 167-170.

564. R. H. Mandl, L. H. Weinstein, J. S. Jacobson, D. C. McCune, and A. E. Hitchcock, in Automation in Analytical Chemistry, Technicon Symposia 1965, Mediad, Inc., New York, 1966, pp. 270-273.

565. R. S. Manly, D. H. Foster, and D. P. Harrington, in Automation in Analytical Chemistry, Technicon Symposia 1966, Vol. I, Mediad, Inc., White Plains, N.Y., 1967, pp. 250-253.

566. R. S. Manly, R. Liberfarb, A. Cormier, and A. O'Brien, in Automation in Analytical Chemistry, Technicon Symposia 1966, Vol. I, Mediad, Inc., White Plains, N.Y., 1967, pp. 254-256.

567. C. K. Mann, Anal. Chem., 29, 1385-1386 (1957).

568. R. Manston, in Automation in Analytical Chemistry, Technicon Symposia 1967, Vol. II, Mediad, Inc., White Plains, N.Y., 1968, pp. 155-158.

569. A. G. Marr and L. Marcus, Anal. Biochem., 2, 576-588 (1961).

570. W. H. Marsh, in Advances in Clinical Chemistry (H. Sobotka and C. P. Stewart, eds.), Vol. 2, Academic Press, New York, 1959, pp. 338-362.

571. W. H. Marsh, in Protides of the Biological Fluids (H. Peeters, ed.), Elsevier, Amsterdam, 1962, pp. 11-26.

572. W. H. Marsh and B. Fingerhut, Clin. Chem., 8, 640-646 (1962).

573. R. T. Marshall and N. Hoover, in Advances in Automated Analysis, Technicon International Congress 1970, Vol. I, Thurman Associates, Miami, 1971, pp. 491-495.

574. R. W. Marsters, Clin. Chem., 8, 91-96 (1962).

575. J. F. Marten, Chem. Ind. (London), 1965, 1365-1367 (1965).

576. J. F. Marten, Proc. Int. Conf. Alcohol Road Traffic, Brit. Med. Ass., London, 3, 226-231 (1962).

577. J. F. Marten, Proc. Meat Ind. Res. Conf. 1966, 126-137 (1966).

578. J. F. Marten and G. Catanzaro, Analyst (London), 91, 42-47 (1966).

579. J. F. Marten and D. R. Grady, in Automation in Analytical Chemistry, Technicon Symposia 1966, Vol. I, Mediad, Inc., White Plains, N.Y., 1967, pp. 546-551.

580. L. E. Martin and C. Harrison, Anal. Biochem., 23, 529-545 (1968).

581. T. J. Martin and C. W. Baird, Med. J. Aust., 1965, I, 463-465 (1965).

582. J. E. Matusik, J. B. Powell, and D. M. Gregory, Clin. Chim. Acta, 39, 15-20 (1972).

583. W. J. McClintock, D. O. Kildsig, W. V. Kessler, and G. S. Banker, J. Pharm. Sci., 57, 1649-1652 (1968).

584. R. L. McCullough, J. K. MacKay, and G. R. Padmanabhan, in Automation in Analytical Chemistry, Technicon Symposia 1967, Vol. I, Mediad, Inc., White Plains, N.Y., 1968, pp. 233-238.

585. W. H. McDaniel, R. N. Hemphill, and W. T. Donaldson, in Automation in Analytical Chemistry, Technicon Symposia 1967, Vol. I, Mediad, Inc., White Plains, N.Y., 1968, pp. 363-367.

586. B. E. McGrew, M. J. F. DuCros, G. W. Stout, and V. H. Falcone, Amer. J. Clin. Pathol., 50, 52-59 (1968).

587. J. M. McKenzie, P. R. Fowler, and V. Fiorica, Anal. Biochem., 16, 139-148 (1966).

588. J. M. McKenzie, P. R. Fowler, and V. Fiorica, in Automation in Analytical Chemistry, Technicon Symposia 1966, Vol. I, Mediad, Inc., White Plains, N.Y., 1967, pp. 45-48.

589. R. D. McNair, Clin. Chem., 10, 648-649 (1964).

590. C. McNeil, C. R. Berrett, W. Rich, S. Gaufin, M. A. Warr, and H. Hutchinson, in Advances in Automated Analysis, Technicon International Congress 1970, Vol. I, Thurman Associates, Miami, 1971, pp. 547-550.

591. D. Melley, A. Little, and R. D. Rothfield, Clin. Chim. Acta, 40, 273-275 (1972).

592. D. E. Mercaldo and E. A. Pizzi, Ann. N.Y. Acad. Sci., 130, 550-557 (1965).

593. R. J. Merrills, Anal. Biochem., 6, 272-282 (1963).

594. R. J. Merrills, in Automation in Analytical Chemistry, Technicon Symposia 1965, Mediad, Inc., New York, 1966, pp. 390-392.

595. R. J. Merrills, Nature (London), 193, 988 (1962).

596. T. P. Michaels, V. J. Greely, W. W. Holl, and L. P. Sinotte, Ann. N.Y. Acad. Sci., 130, 568-574 (1965).

597. T. P. Michaels and L. P. Sinotte, Ann. N.Y. Acad. Sci., 130, 496-503 (1965).

598. W. F. Milbury, V. T. Stack, Jr., and F. L. Doll, in Advances in Automated Analysis, Technicon International Congress 1970, Vol. II, Thurman Associates, Miami, 1971, pp. 299-304.

599. P. J. Milham, Analyst (London), 95, 758-759 (1970).

600. A. B. Miller, E. Hicks, C. M. Jensen, and H. D. Appleton, Clin. Chem., 11, 794-795 (1965).

601. J. P. Mislan and S. Elchuk, in Advances in Automated Analysis, Technicon International Congress 1969, Vol. II, Mediad, Inc., White Plains, N.Y., 1970, pp. 315-320.

602. J. P. Mislan and S. Elchuk, in Automation in Analytical Chemistry, Technicon Symposia 1967, Vol. I, Mediad, Inc., White Plains, N.Y., 1968, pp. 329-332.

603. W. R. Moorehead and E. A. Sasse, Clin. Chem., 16, 285-290 (1970).

604. J.-R. Mor, A. Zimmerli, and A. Fiechter, Anal. Biochem., 52, 614-624 (1973).

605. F. B. Moreland, Amer. J. Clin. Pathol., 43, 168 (1965).

606. G. B. Morgan, E. C. Tabor, C. Golden, and H. Clements, in Automation in Analytical Chemistry, Technicon Symposia 1966, Vol. I, Mediad, Inc., White Plains, N.Y., 1967, pp. 534-541.

607. S. Morgenstern, L. Chaparian, D. Vlastelica, and A. Kiederer, in Advances in Automated Analysis, Technicon International Congress 1970, Vol. I, Thurman Associates, Miami, 1971, pp. 85-90.

608. S. Morgenstern, R. Flor, G. Kessler, and B. Klein, Anal. Biochem., 13, 149-161 (1965).

609. R. Moss and E. V. Browett, Analyst (London), 91, 428-438 (1966).

610. R. Moss and E. V. Browett, in Automation in Analytical Chemistry, Technicon Symposia 1965, Mediad, Inc., New York, 1966, pp. 285-290.

611. G. G. Muir and M. Ryan, Clin. Chem., 17, 1007-1009 (1971).

612. G. G. Muir, M. Ryan, and D. U. Conaill, J. Clin. Pathol., 23, 492-498 (1970).

613. G. G. Muir, D. UaConnaill, and M. Ryan, Steroids, 13, 719-722 (1969).

614. O. H. Muller, J. Amer. Chem. Soc., 69, 2992-2997 (1947).

615. R. H. Muller, Anal. Chem., 30, (1), 53A-54A, 56A (1958).

616. K. W. Mundry, in Automation in Analytical Chemistry, Technicon Symposia 1965, Mediad, Inc., New York, 1966, pp. 612-614.

617. P. D. Murphy and G. K. Summer, in Advances in Automated Analysis, Technicon International Congress 1970, Vol. I, Thurman Associates, Miami, 1971, pp. 513-516.

618. A. L. Nagler, C. Caban, A. Hughes, and T. Keenan, J. Lab. Clin. Med., 68, 131-136 (1966).

619. R. M. Nalbandian, B. M. Nichols, F. R. Camp, Jr., J. M. Lusher, N. F. Conte, R. L. Henry, and P. L. Wolf, Clin. Chem., 17, 1033-1037 (1971).

620. H. N. Naumann, A. M. Olsen, and J. M. Young, Clin. Chem., 7, 70-74 (1961).

621. W. E. Neeley, Clin. Chem., 18, 509-515 (1972).

622. W. E. Neeley, R. L. Cechner, R. L. Martin, and G. Marshall, Amer. J. Clin. Pathol., 56, 493-499 (1971).

623. W. E. Neeley, G. E. Goldman, and C. A. Cupas, Clin. Chem., 18, 1350-1354 (1972).

623a. W. E. Neeley, S. Wardlaw, and M. E. T. Swinnen, Clin. Chem., 20, 78-80 (1974).

624. M. G. Nelson and A. Lamont, J. Clin. Pathol., 14, 448-450 (1961).

625. D. B. Nevius and G. F. Lanchantin, Clin. Chem., 11, 633-640 (1965).

626. P. J. Niebergall and J. E. Goyan, J. Pharm. Sci., 52, 29-33 (1963).

627. J. A. Nisbet and E. Simpson, Clin. Chim. Acta, 39, 339-350 (1972).

628. H. H. Nishi and A. Rhodes, in Automation in Analytical Chemistry, Technicon Symposia 1965, Mediad, Inc., New York, 1966, pp. 321-323.

629. R. P. Noble and F. M. Campbell, Clin. Chem., 16, 166-170 (1970).

630. C. D. Nordschow and A. R. Tammes, Anal. Chem., 40, 465-466 (1968).

631. L. Nyberg, J. Pharm. Pharmacol., 22, 500-506 (1970).

632. J. E. O'Brien, in Automation in Analytical Chemistry, Technicon Symposia 1966, Vol. I, Mediad, Inc., White Plains, N.Y., 1967, pp. 591-594.

633. J. E. O'Brien and J. Fiore, Wastes Eng., 33, 128-131, 147 (1962).

634. W. S. Oleniacz, M. A. Pisano, and M. H. Rosenfield, in Automation in Analytical Chemistry, Technicon Symposia 1966, Vol. I, Mediad, Inc., White Plains, N.Y., 1967, pp. 523-525.

635. R. T. Oliver, G. F. Lenz, and W. P. Frederick, in Advances in Automated Analysis, Technicon International Congress 1969, Vol. II, Mediad, Inc., White Plains, N.Y., 1970, pp. 309-314.

636. D. E. Ott and F. A. Gunther, J. Ass. Offic. Anal. Chem., 49, 662-669 (1966).

637. D. E. Ott and F. A. Gunther, J. Ass. Offic. Anal. Chem., 49, 669-674 (1966).

638. D. E. Ott and F. A. Gunther, J. Ass. Offic. Anal. Chem., 51, 697-708 (1968).

639. D. E. Ott, M. Ittig, and H. O. Friestad, J. Ass. Offic. Anal. Chem., 54, 160-164 (1971).

640. C. A. Owen, Jr., F. T. Maher, W. F. McGuckin, and J. Issacson, Biomed. Sci. Instrum., 1, 5-10 (1963).

641. J. A. Owen, Clin. Chim. Acta, 14, 426-428 (1966).

642. J. A. Owen, Clin. Chim. Acta, 29, 89-91 (1970).

643. J. F. Pagano, T. A. Haney, and J. R. Gerke, Ann. N.Y. Acad. Sci., 93, 644-648 (1962).

644. D. W. Palmer and T. Peters, Jr., Clin. Chem., 15, 891-901 (1969).

645. H. L. Pardue, in Advances in Analytical Chemistry and Instrumentation (C. N. Reilley and F. W. McLafferty, eds.), Vol. 7, Wiley (Interscience), New York, 1968, pp. 141-207.

646. H. L. Pardue and S. N. Deming, Anal. Chem., 41, 986-989 (1969).

647. H. L. Pardue and P. A. Rodriguez, Anal. Chem., 39, 901-907 (1967).

648. S. Passen and W. Gennaro, Amer. J. Clin. Pathol., 46, 69-81 (1966).

649. S. Passen and R. Von Saleski, Amer. J. Clin. Pathol., 51, 166-176 (1969).

650. B. J. Payne, S. M. Free, Jr., and G. W. Lucyszyn, in Advances in Automated Analysis, Technicon International Congress 1969, Vol. III, Mediad, Inc., White Plains, N.Y., 1970, pp. 127-131.

651. O. Pelletier and R. Brassard, J. Ass. Offic. Anal. Chem., 54, 1164-1167 (1971).

652. G. Pennacchia, V. G. Bethune, M. Fleisher, and M. K. Schwartz, Clin. Chem., 17, 339-340 (1971).

653. E. I. Pentz, in Advances in Automated Analysis, Technicon International Congress 1969, Vol. I, Mediad, Inc., White Plains, N.Y., 1970, pp. 111-116.

654. E. I. Pentz, Anal. Biochem., 8, 328-336 (1964).

655. E. I. Pentz, Anal. Biochem., 27, 333-342 (1969).

656. J. C. A. Peoples, S. L. Adler, and C. V. Johnson, in Advances in Automated Analysis, Technicon International Congress 1969, Vol. I, Mediad, Inc., White Plains, N.Y., 1970, pp. 293-295.

657. G. A. Persson, Int. J. Air Water Pollut., 10, 845-852 (1966).

658. J. D. Peuler and P. G. Passon, Anal. Biochem., 52, 574-583 (1973).

659. M. A. Pinnegar, in Automation in Analytical Chemistry, Technicon Symposia 1965, Mediad, Inc., New York, 1966, pp. 80-83.

660. H. C. Pitot and N. Pries, Anal. Biochem., 9, 454-466 (1964).

661. H. C. Pitot, N. Pries, M. Poirier, and A. Cutler, in Automation in Analytical Chemistry, Technicon Symposia 1965, Mediad, Inc., New York, 1966, pp. 555-558.

662. H. C. Pitot, N. Wratten, and M. Poirier, Anal. Biochem., 22, 359-373 (1968).

663. R. Place and F. Hardy, in Automation in Analytical Chemistry, Technicon Symposia 1966, Vol. I, Mediad, Inc., White Plains, N.Y., 1967, pp. 649-651.

664. T. B. Platt, H. Weisblatt, and L. Guevrekian, Ann. N.Y. Acad. Sci., 153, 571-581 (1968).

665. B. J. Poletti, J. F. Zack, Jr., and T. J. Mueller, Amer. J. Clin. Pathol., 53, 731-738 (1970).

666. A. Pollard and C. B. Waldron, in Automation in Analytical Chemistry, Technicon Symposia 1966, Vol. I, Mediad, Inc., White Plains, N.Y., 1967, pp. 49-58.

667. A. C. Pollard, E. S. Garnett, and C. E. Webber, in Automation in Analytical Chemistry, Technicon Symposia 1965, Mediad, Inc., New York, 1966, pp. 387-389.

668. A. C. Pollard, E. S. Garnett, and C. Webber, Clin. Chem., 11, 959-967 (1965).

669. F. H. Pollard, G. Nickless, D. E. Rodgers, and M. T. Rothwell, J. Chromatogr., 17, 157-167 (1965).

670. D. G. Porter and R. Sawyer, Analyst (London), 97, 569-575 (1972).

671. S. Posen, D. J. Birkett, R. A. J. Conyers, C. J. Cornish, and F. C. Neale, in Automation in Analytical Chemistry, Technicon Symposia 1967, Vol. I, Mediad, Inc., White Plains, N.Y., 1968, pp. 583-585.

672. F. J. N. Powell, in Automation in Analytical Chemistry, Technicon Symposia 1967, Vol. II, Mediad, Inc., White Plains, N.Y., 1968, pp. 103-104.

673. J. B. Powell, C. E. Emery, and G. A. Peyton, Clin. Chem., 18, 1318-1322 (1972).

674. Practical Automation for the Clinical Laboratory (W. L. White, M. M. Erickson, and S. C. Stevens, eds.), C. V. Mosby Co., St. Louis, 1968, pp. 36-186, 195-198, 309-343.

675. J. R. Prall, in Advances in Automated Analysis, Technicon International Congress 1969, Vol. II, Mediad, Inc., White Plains, N.Y., 1970, pp. 297-303.

676. J. R. Prall, in Advances in Automated Analysis, Technicon International Congress 1970, Vol. II, Thurman Associates, Miami, 1971, pp. 395-398.

677. J. R. Prall, in Advances in Automated Analysis, Technicon International Congress 1970, Vol. II, Thurman Associates, Miami, 1971, pp. 399-402.

678. G. C. Prescott, in Automation in Analytical Chemistry, Technicon Symposia 1965, Mediad, Inc., New York, 1966, pp. 38-41.

679. G. C. Prescott, J. Pharm. Sci., 55, 423-425 (1966).

680. Principles of Operation, Basic AutoAnalyzer System, Technicon Corporation, Tarrytown, N.Y., 1969.

681. H. N. Rasmussen and J. R. Nielsen, Anal. Biochem., 50, 648-651 (1972).

682. R. L. Rebertus, R. J. Cappell, and G. W. Bond, Anal. Chem., 30, 1825-1827 (1958).

683. D. J. Reed, M. Kennedy, and W. Lapsa, in Advances in Automated Analysis, Technicon International Congress 1970, Vol. I, Thurman Associates, Miami, 1971, pp. 359-363.

684. R. H. P. Reid and L. Wise, in Automation in Analytical Chemistry, Technicon Symposia 1967, Vol. II, Mediad, Inc., White Plains, N.Y., 1968, pp. 159-165.

685. B. Rexen and B. Christensen, in Automation in Analytical Chemistry, Technicon Symposia 1966, Vol. II, Mediad, Inc., White Plains, N.Y., 1967, pp. 435-438.

686. R. C. Robbins, Clin. Chem., 15, 56-60 (1969).

687. D. L. Robertson, F. Matsui, and W. N. French, Can. J. Pharm. Sci.,
     7, 47-50 (1972).

688. J. A. S. Rokos, S. B. Rosalki, and D. Tarlow, Clin. Chem., 18,
     193-198 (1972).

689. D. Roland, G. R. Cooper, and R. F. Witter, J. Amer. Oil Chem. Soc.,
     43, 649-651 (1966).

690. D. B. Roodyn, in Automation in Analytical Chemistry, Technicon
     Symposia 1965, Mediad, Inc., New York, 1966, pp. 593-594.

691. D. B. Roodyn, in Automation in Analytical Chemistry, Technicon
     Symposia 1967, Vol. II, Mediad, Inc., White Plains, N.Y., 1968,
     pp. 233-237.

692. D. B. Roodyn, Nature (London), 206, 1226-1228 (1965).

693. D. B. Roodyn and N. G. Maroudas, Anal. Biochem., 24, 496-505
     (1968).

694. C. F. M. Rose, Lab. Pract., 12, 57-61 (1963).

695. W. T. Roubal, Anal. Chem., 37, 440-442 (1965).

696. W. T. Roubal, J. Chem. Educ., 45, 439 (1968).

697. W. T. Roubal and A. T. Tappel, Anal. Biochem., 9, 211-216 (1964).

698. H. E. Roudebush, Ann. N.Y. Acad. Sci., 130, 582-588 (1965).

699. M. E. Royer and H. Ko, Anal. Biochem., 29, 405-416 (1969).

700. G. Rudnik, in Automation in Analytical Chemistry, Technicon
     Symposia 1965, Mediad, Inc., New York, 1966, pp. 220-223.

701. A. H. Runck and C. R. Valeri, in Advances in Automated Analysis,
     Technicon International Congress 1970, Vol. I, Thurman Associates,
     Miami, 1971, pp. 477-479.

702. R. L. Rush, L. Leon, and J. Turrell, in Advances in Automated
     Analysis, Technicon International Congress 1970, Vol. I, Thurman
     Associates, Miami, 1971, pp. 503-507.

703. F. M. Russo-Alesi, Ann. N.Y. Acad. Sci., 153, 511-524 (1968).

704. F. M. Russo-Alesi and A. J. Khoury, in Automation in Analytical
     Chemistry, Technicon Symposia 1967, Vol. I, Mediad, Inc., White
     Plains, N.Y., 1968, pp. 491-495.

705. F. M. Russo-Alesi, C. Sherman, J. M. Kelly, and A. Ferrari,
     Ann. N.Y. Acad. Sci., 87, 822-829 (1960).

706. E. R. Rutter, Clin. Chem., 18, 616-620 (1972).

707. J. Ruzicka and C. G. Lamm, in Automation in Analytical Chemistry, Technicon Symposia 1967, Vol. II, Mediad, Inc., White Plains, N.Y., 1968, pp. 315-319.

708. J. Ruzicka and C. G. Lamm, Talanta, 15, 689-697 (1968).

709. J. Ruzicka and M. Williams, Talanta, 12, 967-970 (1965).

710. J. A. Ryan, E. McGonigle, and J. M. Konieczny, Anal. Chim. Acta, 55, 83-87 (1971).

711. W. T. Ryan and S. Morgenstern, in Advances in Automated Analysis, Technicon International Congress 1970, Vol. II, Thurman Associates, Miami, 1971, pp. 433-435.

712. A. L. Ryland, W. P. Pickhardt, and C. D. Lewis, 1964 Technicon International Symposium, Interim Reprint, New York, "Automation Techniques in Analytical Research."

713. O. Samuelson, L. Larsson, and O. Ramnas, in Automation in Analytical Chemistry, Technicon Symposia 1965, Mediad, Inc., New York, 1966, pp. 169-173.

714. O. Samuelson and B. Swenson, Anal. Chim. Acta, 28, 426-432 (1963).

715. R. F. Sankuer, T. Geist, and F. Henry, in Automation in Analytical Chemistry, Technicon Symposia 1966, Vol. I, Mediad, Inc., White Plains, N.Y., 1967, pp. 613-616.

716. M. Sansur, A. Buccafuri, and S. Morgenstern, J. Ass. Offic. Anal. Chem., 55, 880-887 (1972).

717. F. Santacana, in Automation in Analytical Chemistry, Technicon Symposia 1967, Vol. I, Mediad, Inc., White Plains, N.Y., 1968, pp. 339-345.

718. F. Santacana and S. Mitzner, in Automation in Analytical Chemistry, Technicon Symposia 1966, Vol. I, Mediad, Inc., White Plains, N.Y., 1967, pp. 622-626.

719. E. A. Sasse and W. R. Moorehead, in Advances in Automated Analysis, Technicon International Congress 1969, Vol. I, Mediad, Inc., White Plains, N.Y., 1970, pp. 61-67.

720. A. M. Saunders, Clin. Chem., 18, 783-788 (1972).

721. A. M. Saunders, W. Groner, and J. Kusnetz, in Advances in Automated Analysis, Technicon International Congress 1970, Vol. I, Thurman Associates, Miami, 1971, pp. 453-459.

722. J. Savory, M. G. Heintges, L. M. Killingsworth, and J. M. Potter, Clin. Chem., 18, 37-42 (1972).

723. R. Sawyer, in Automation in Analytical Chemistry, Technicon Symposia 1967, Vol. I, Mediad, Inc., White Plains, N.Y., 1968, pp. 227-231.

724. R. Sawyer and E. J. Dixon, Analyst (London), 93, 669-679 (1968).

725. R. Sawyer and E. J. Dixon, Analyst (London), 93, 680-687 (1968).

726. R. Sawyer and E. J. Dixon, in Automation in Analytical Chemistry, Technicon Symposia 1966, Vol. II, Mediad, Inc., White Plains, N.Y., 1967, pp. 111-116.

727. R. Sawyer, E. J. Dixon, and E. Johnston, Analyst (London), 94, 1010-1020 (1969).

728. E. Scarano, M. G. Bonicelli, and M. Forina, Anal. Chem., 42, 1470-1472 (1970).

729. G. D. Schaiberger and A. Ferrari, Ann. N.Y. Acad. Sci., 87, 890-893 (1960).

730. R. A. Scheidt, V. A. Nelson, and J. B. Levine, in Automation in Analytical Chemistry, Technicon Symposia 1965, Mediad, Inc., New York, 1966, pp. 563-568.

731. G. Scheuerbrandt, Anal. Biochem., 13, 475-482 (1965).

732. G. Schill, Acta Pharm. Suecica, 2, 13-46 (1965).

733. P. H. Scholes and C. Thulbourne, Analyst (London), 88, 702-712 (1963).

734. P. H. Scholes and C. Thulbourne, Analyst (London), 89, 466-474 (1964).

735. W. A. Schroeder, R. T. Jones, J. Cormick, and K. McCalla, Anal. Chem., 34, 1570-1575 (1962).

736. W. A. Schroeder and B. Robberson, Anal. Chem., 37, 1583-1585 (1965).

737. W. C. Schroeder, in Automation in Analytical Chemistry, Technicon Symposia 1966, Vol. I, Mediad, Inc., White Plains, N.Y., 1967, pp. 456-457.

738. A. L. Schroeter, H. F. Taswell, R. R. Kierland, and M. A. Sweatt, Amer. J. Clin. Pathol., 56, 43-49 (1971).

739. A. L. Schroeter, H. F. Taswell, and M. A. Sweatt, in Advances in Automated Analysis, Technicon International Congress 1969, Vol. I, Mediad, Inc., White Plains, N.Y., 1970, pp. 265-268.

740. L. C. Schroeter and W. E. Hamlin, J. Pharm. Sci., 52, 811-812 (1963).

741. H. Schuel and R. Schuel, Anal. Biochem., 20, 86-93 (1967).

742. H. Schuel, R. Schuel, and N. J. Unakar, Anal. Biochem., 25, 146-163 (1968).

743. W. D. Schultze, in Advances in Automated Analysis, Technicon International Congress 1970, Vol. II, Thurman Associates, Miami, 1971, pp. 87-92.

744. K. F. Schunk, Ann. N.Y. Acad. Sci., 87, 924-933 (1960).

745. L. Schutte, J. Chromatogr., 72, 303-309 (1972).

746. M. K. Schwartz, in Automation in Analytical Chemistry, Technicon Symposia 1967, Vol. I, Mediad, Inc., White Plains, N.Y., 1968, pp. 587-592.

747. M. K. Schwartz and O. Bodansky, in Automation in Analytical Chemistry, Technicon Symposia 1966, Vol. I, Mediad, Inc., White Plains, N.Y., 1967, pp. 489-493.

748. M. K. Schwartz and O. Bodansky, in Methods of Biochemical Analysis (D. Glick, ed.), Vol. 11, Wiley (Interscience), New York, 1963, pp. 211-246.

749. M. K. Schwartz and O. Bodansky, in Methods of Biochemical Analysis (D. Glick, ed.), Vol. 16, Wiley (Interscience), New York, 1968, pp. 183-218.

750. M. K. Schwartz, G. Kessler, and O. Bodansky, Ann. N.Y. Acad. Sci., 87, 616-628 (1960).

751. M. K. Schwartz, G. Kessler, and O. Bodansky, J. Biol. Chem., 236, 1207-1211 (1961).

752. H. D. Scobell, H. Tai, and J. B. Hill, in Advances in Automated Analysis, Technicon International Congress 1969, Vol. II, Mediad, Inc., White Plains, N.Y., 1970, pp. 79-82.

753. W. S. Sebborn, Analyst (London), 94, 324-329 (1969).

754. E. Seifter, D. Kambosos, and A. Chanas, in Advances in Automated Analysis, Technicon International Congress 1969, Vol. I, Mediad, Inc., White Plains, N.Y., 1970, pp. 121-125.

755. E. Seifter, D. Kambosos, and A. Chanas, in Advances in Automated Analysis, Technicon International Congress 1970, Vol. I, Thurman Associates, Miami, 1971, pp. 509-511.

756. W. M. Sergy, in Automation in Analytical Chemistry, Technicon Symposia 1965, Mediad, Inc., New York, 1966, pp. 123-130.

757. R. Shapira and A. M. Wilson, Anal. Chem., 38, 1803 (1966).

758. W. H. Shaw and R. E. Duncombe, Ann. N.Y. Acad. Sci., 130, 647-656 (1965).

759. W. H. C. Shaw and R. E. Duncombe, Analyst (London), 88, 694-701 (1963).

760. W. H. C. Shaw and I. Fortune, Analyst (London), 87, 187-196 (1962).

761. J. M. Shepp, J. P. Messerly, and J. Bomstein, Anal. Biochem., 8, 122-124 (1964).

762. O. Siggaard-Andersen and D. Oliver, Scand. J. Clin. Lab. Invest., 21, 92-94 (1968).

763. R. Simonson, in Automation in Analytical Chemistry, Technicon Symposia 1967, Vol. II, Mediad, Inc., White Plains, N.Y., 1968, pp. 291-294.

764. F. M. Singer, A. Borman, and L. J. Lerner, Ann. N.Y. Acad. Sci., 102, 118-126 (1962).

765. J. A. de S. Siriwardene, R. A. Evans, A. J. Thomas, and R. F. E. Axford, in Automation in Analytical Chemistry, Technicon Symposia 1965, Mediad, Inc., New York, 1966, pp. 144-147.

766. L. T. Skeggs, U. S. patent 2,797,149, June 25, 1957; through Chem. Abstr., 51, 12573d (1957), "Apparatus for Determination of Crystalloid Constituents of a Liquid that Contains also Noncrystalloid Substances."

767. L. T. Skeggs, Jr., Amer. J. Clin. Pathol., 28, 311-322 (1957).

768. L. T. Skeggs, Jr., Amer. J. Clin. Pathol., 33, 181-185 (1960).

769. L. T. Skeggs, Anal. Chem., 38, (6), 31A-34A, 36A, 38A, 40A, 42A-44A (1966).

770. L. T. Skeggs, Jr., Ann. N.Y. Acad. Sci., 87, 650-657 (1960).

771. L. T. Skeggs, Jr., U. S. patent 3,230,048 (Cl. 23-253), Jan. 18, 1966; Appl. Aug. 17, 1962; 8 pp.; through Chem. Abstr., 64, 8929b (1966), "Chromatography Analysis Apparatus."

772. L. T. Skeggs, Jr., Clin. Chem., 2, 241 (1956).

773. L. T. Skeggs, Jr., in Standard Methods of Clinical Chemistry (S. Meites, ed.), Vol. 5, Academic Press, New York, 1965, pp. 31-42.

774. L. T. Skeggs, Jr., Tech. Bull. Regist. Med. Technol. Amer. Soc. Clin. Pathol., 30, 1-5 (1960).

775. L. T. Skeggs, Jr., and H. Hochstrasser, Clin. Chem., 10, 918-936 (1964).

776. J. M. Skinner, in Automation in Analytical Chemistry, Technicon Symposia 1967, Vol. II, Mediad, Inc., White Plains, N.Y., 1968, pp. 327-332.

777. J. M. Skinner and A. C. Docherty, Talanta, 14, 1393-1401 (1967).

778. K. G. Sloman and M. Panio, in Advances in Automated Analysis, Technicon International Congress 1969, Vol. II, Mediad, Inc., White Plains, N.Y., 1970, pp. 83-85.

779. R. V. Smith, L. L. Ciaccio, and R. L. Lipchus, in Automation in Analytical Chemistry, Technicon Symposia 1965, Mediad, Inc., New York, 1966, pp. 57-60.

780. R. V. Smith, L. L. Ciaccio, and R. L. Lipchus, J. Agr. Food Chem., 15, 408-411 (1967).

781. W. J. Smythe, M. H. Shamos, S. Morgenstern, and L. T. Skeggs, in Automation in Analytical Chemistry, Technicon Symposia 1967, Vol. I, Mediad, Inc., White Plains, N.Y., 1968, pp. 105-113.

782. P. Z. Sobocinski and R. P. McDevitt, Clin. Chem., 18, 487 (1972).

783. A. Sodergren, Analyst (London), 91, 113-118 (1966).

784. M. Sparagana, L. Kucera, G. Phillips, and C. Hoffman, in Advances in Automated Analysis, Technicon International Congress 1969, Vol. II, Mediad, Inc., White Plains, N.Y., 1970, pp. 201-205.

785. M. Sparagana, L. Kucera, G. Phillips, and C. Hoffmann, Steroids, 15, 353-371 (1970).

786. D. Stansfield and D. N. Rossington, in Automation in Analytical Chemistry, Technicon Symposia 1965, Mediad, Inc., New York, 1966, pp. 532-534.

787. R. E. Stanton and A. J. McDonald, Analyst (London), 88, 608-613 (1963).

788. R. E. Stanton and A. J. McDonald, Chem. Ind. (London), 1961, (35), 1406-1407 (1961).

789. T. M. Steele and M. C. Mansdorfer, Amer. J. Clin. Pathol., 53, 116-120 (1970).

790. L. D. Stegink, in Advances in Automated Analysis, Technicon International Congress 1970, Vol. I, Thurman Associates, Miami, 1971, pp. 591-594.

791. H. H. Stein, Anal. Biochem., 13, 305-313 (1965).

792. H. H. Stein, in Automation in Analytical Chemistry, Technicon Symposia 1965, Mediad, Inc., New York, 1966, pp. 45-49.

793. H. H. Stein, A. J. Glasky, and W. R. Roderick, Ann. N.Y. Acad. Sci., 130, 751-760 (1965).

794. C. E. Stevenson, L. D. Bechtel, and L. J. Coursen, in Advances in Automated Analysis, Technicon International Congress 1969, Vol. II, Mediad, Inc., White Plains, N.Y., 1970, pp. 251-256.

795. C. E. Stevenson and I. Comer, in Automation in Analytical Chemistry, Technicon Symposia 1967, Vol. I, Mediad, Inc., White Plains, N.Y., 1968, pp. 483-486.

796. C. E. Stevenson and I. Comer, J. Pharm. Sci., 57, 1227-1230 (1968).

797. K. K. Stewart, Anal. Biochem., 51, 11-18 (1973).

798. W. S. Stewart, J. B. Hall, R. T. Mullen, and H. R. Skeggs, in Advances in Automated Analysis, Technicon International Congress 1970, Vol. II, Thurman Associates, Miami, 1971, pp. 437-446.

799. P. B. Stockwell and R. Sawyer, Anal. Chem., 42, 1136-1141 (1970).

800. H. W. Stowe and M. F. Pelletier, in Automation in Analytical Chemistry, Technicon Symposia 1967, Vol. I, Mediad, Inc., White Plains, N.Y., 1968, pp. 431-433.

801. P. E. Strandjord and K. J. Clayson, J. Lab. Clin. Med., 67, 131-143 (1966).

802. P. E. Strandjord and K. J. Clayson, J. Lab. Clin. Med., 67, 154-170 (1966).

803. H. S. Strickler, S. S. Holt, H. F. Acevedo, E. Saier, and R. C. Grauer, Steroids, 9, 193-216 (1967).

804. H. S. Strickler, S. S. Holt, R. C. Grauer, and J. Gilmore, in Automation in Analytical Chemistry, Technicon Symposia 1966, Vol. I, Mediad, Inc., White Plains, N.Y., 1967, pp. 70-74.

805. H. S. Strickler, J. M. Petty, and P. J. Stanchak, Clin. Chem., 17, 1186-1190 (1971).

806. H. S. Strickler, E. L. Saier, and R. C. Grauer, in Automation in Analytical Chemistry, Technicon Symposia 1965, Mediad, Inc., New York, 1966, pp. 368-371.

807. H. S. Strickler and P. J. Stanchak, Clin. Chem., 15, 137-153 (1969).

808. H. S. Strickler, P. J. Stanchak, and J. J. Maydak, Anal. Chem., 42, 1576-1578 (1970).

809. W. A. Stuart, Analyst (London), 91, 208-210 (1966).

810. P. Sturgeon, Amer. J. Clin. Pathol., 52, 50-56 (1969).

811. P. Sturgeon, M. DuCros, D. McQuiston, and W. Smythe, in Automation in Analytical Chemistry, Technicon Symposia 1965, Mediad, Inc., New York, 1966, pp. 515-524.

812. P. Sturgeon and D. T. McQuiston, Amer. J. Clin. Pathol., 43, 517-531 (1965).

813. P. Sturgeon and D. McQuiston, in Automation in Analytical Chemistry, Technicon Symposia 1966, Vol. I, Mediad, Inc., White Plains, N.Y., 1967, pp. 122-125.

814. E. Suba, in Advances in Automated Analysis, Technicon International Congress 1970, Vol. I, Thurman Associates, Miami, 1971, pp. 337-342.

815. R. B. Summers, R. B. Mefferd, Jr., and H. R. Littleton, 1964 Technicon International Symposium, Interim Reprint, New York, "Pressurized Automated Systems for Micro-Analysis--With an Application to the Determination of Serum Copper, Magnesium, and Iron."

816. R. M. Summers, Anal. Chem., 32, 1903-1904 (1960); see also Anal. Chem., 33, 872 (1961).

817. A. H. Sutton and I. A. R. Duthie, in Automation in Analytical Chemistry, Technicon Symposia 1967, Vol. II, Mediad, Inc., White Plains, N.Y., 1968, pp. 333-338.

818. R. A. Syed, in Automation in Analytical Chemistry, Technicon Symposia 1967, Vol. I, Mediad, Inc., White Plains, N.Y., 1968, pp. 571-577.

819. A. R. Tammes, R. M. Steadman, and D. T. Hunter, Amer. J. Clin. Pathol., 54, 231-234 (1970).

820. I. K. Tan and T. P. Whitehead, Clin. Chem., 15, 467-478 (1969).

821. A. L. Tappel and C. Beck, in Automation in Analytical Chemistry, Technicon Symposia 1965, Mediad, Inc., New York, 1966, pp. 559-562.

822. A. L. Tappel and C. Beck, in Automation in Analytical Chemistry, Technicon Symposia 1967, Vol. I, Mediad, Inc., White Plains, N.Y., 1968, pp. 593-598.

823. A. F. Taylor and M. E. Northmore, in Automation in Analytical Chemistry, Technicon Symposia 1967, Vol. II, Mediad, Inc., White Plains, N.Y., 1968, pp. 263-267.

824. I. E. Taylor and M. M. Marsh, Amer. J. Clin. Pathol., 32, 393-396 (1959).

825. I. E. Taylor and M. M. Marsh, Ann. N.Y. Acad. Sci., 87, 775-781 (1960).

826. Technicon AutoAnalyzer Bibliography 1957/1967, Technicon Corporation, Tarrytown, N.Y., 1968.

827. Technicon AutoAnalyzer Pharmaceutical Bibliography, Technicon Industrial Systems, Tarrytown, N.Y., 1971.

828. A. M. Tenny, in Automation in Analytical Chemistry, Technicon Symposia 1966, Vol. I, Mediad, Inc., White Plains, N.Y., 1967, pp. 580-583.

829. A. C. Terranova, J. G. Pomonis, R. F. Severson, and P. A. Hermes, in Automation in Analytical Chemistry, Technicon Symposia 1967, Vol. I, Mediad, Inc., White Plains, N.Y., 1968, pp. 501-505.

830. R. E. Thiers, Clin. Chem., 10, 651 (1964).

830a. R. E. Thiers, in Clinical Chemistry (R. J. Henry, D. C. Cannon, and J. W. Winkelman, eds.), Chap. 10, Harper & Row, Hagerstown, Md., 1974.

831. R. E. Thiers, J. Bryan, and K. Oglesby, Clin. Chem., 12, 120-136 (1966).

832. R. E. Thiers, R. R. Cole, and W. J. Kirsch, Clin. Chem., 13, 451-467 (1967).

833. R. E. Thiers, W. J. Kirsch, and R. R. Cole, in Automation in Analytical Chemistry, Technicon Symposia 1966, Vol. I, Mediad, Inc., White Plains, N.Y., 1967, pp. 37-44.

834. R. E. Thiers, J. Meyn, and R. F. Wildermann, Clin. Chem., 16, 832-839 (1970).

835. R. E. Thiers and K. M. Oglesby, Clin. Chem., 10, 246-257 (1964).

835a. R. E. Thiers, A. H. Reed, and K. Delander, Clin. Chem., 17, 42-48 (1971).

836. A. J. Thomas, R. A. Evans, J. A. de S. Siriwardene, and A. J. Robins, Biochem. J., 99, 5C-7C (1966).

837. G. Thomas, in Automation in Analytical Chemistry, Technicon Symposia 1965, Mediad, Inc., New York, 1966, pp. 148-152.

838. H. I. Thompson and G. A. Rechnitz, Anal. Chem., 44, 300-305 (1972).

839. J. F. Thompson, in Automation in Analytical Chemistry, Technicon Symposia 1966, Vol. I, Mediad, Inc., White Plains, N.Y., 1967, pp. 458-461.

840. J. F. Thompson, C. J. Morris, and R. C. Hodson, in Automation in Analytical Chemistry, Technicon Symposia 1965, Mediad, Inc., New York, 1966, pp. 84-93.

841. N. W. Tietz and A. Green, Clin. Chim. Acta, 9, 392-399 (1964).

842. N. W. Tietz and A. Green, Clin. Chim. Acta, 14, 566-567 (1966).

843. E. O. Titus, W. W. Holland, E. A. Brown, and W. M. Hart, Jr., Anal. Biochem., 54, 40-46 (1973).

844. R. Tkachuk, Can. J. Chem., 40, 2348-2356 (1962).

845. J. R. Todd and J. H. Byars, in Automation in Analytical Chemistry, Technicon Symposia 1966, Vol. I, Mediad, Inc., White Plains, N.Y., 1967, pp. 140-144.

846. Y. Torud, Pharm. Acta Helv., 46, 248-256 (1971).

847. K. Tsuji, Ann. N.Y. Acad. Sci., 153, 446-455 (1968).

848. J. H. Tufekjian, W. W. Holl, T. P. Michaels, and L. P. Sinotte, Ann. N.Y. Acad. Sci., 130, 621-626 (1965).

849. D. E. Uhl, E. B. Lancaster, and C. Vojnovich, Anal. Chem., 43, 990-994 (1971).

850. T. Urbanyi and M. C. H. Lin, in Advances in Automated Analysis, Technicon International Congress 1970, Vol. II, Thurman Associates, Miami, 1971, pp. 253-257.

851. T. Urbanyi and M. C. H. Lin, J. Pharm. Sci., 60, 755-758 (1971).

852. T. Urbanyi and A. O'Connell, Anal. Chem., 44, 565-570 (1972).

853. T. Urbanyi and H. Stober, J. Ass. Offic. Anal. Chem., 55, 180-184 (1972).

854. J. P. Ussary and C. W. Gehrke, in Advances in Automated Analysis, Technicon International Congress 1969, Vol. II, Mediad, Inc., White Plains, N.Y., 1970, pp. 89-94.

855. J. P. Ussary and C. W. Gehrke, J. Ass. Offic. Agr. Chem., 48, 865-868 (1965).

856. L. Valentini, in Advances in Automated Analysis, Technicon International Congress 1969, Vol. II, Mediad, Inc., White Plains, N.Y., 1970, pp. 87-88.

857. L. Valentini, in Automation in Analytical Chemistry, Technicon Symposia 1966, Vol. II, Mediad, Inc., White Plains, N.Y., 1967, pp. 27-29.

858. H. Van Belle, in Automation in Analytical Chemistry, Technicon Symposia 1967, Vol. II, Mediad, Inc., White Plains, N.Y., 1968, pp. 275-279.

859. A. Vandermeers, H. Lclotte, and J. Christophe, Anal. Biochem., 42, 437-445 (1971).

860. A. Vandermeers, M. C. Vandermeers-Piret, and J. Christophe, Clin. Chim. Acta, 37, 471-476 (1972).

861. P. A. Van Dreal, M. E. Beck, and R. Hansen, Clin. Chem., 15, 812 (1969).

862. K. Van Dyke, R. Stitzel, T. McClellan, and C. Szustkiewicz, in Advances in Automated Analysis, Technicon International Congress 1969, Vol. I, Mediad, Inc., White Plains, N.Y., 1970, pp. 47-53.

863. K. Van Dyke, R. Stitzel, T. McClellan, and C. Szustkiewicz, Clin. Chem., 15, 3-14 (1969).

864. K. Van Dyke and C. Szustkiewicz, Anal. Biochem., 23, 109-116 (1968).

865. K. Van Dyke and C. Szustkiewicz, in Automation in Analytical Chemistry, Technicon Symposia 1967, Vol. I, Mediad, Inc., White Plains, N.Y., 1968, pp. 543-548.

866. R. Vargues, Ann. N.Y. Acad. Sci., 130, 819-826 (1965).

867. R. Vargues, in Automation in Analytical Chemistry, Technicon Symposia 1965, Mediad, Inc., New York, 1966, pp. 507-514.

868. R. Vargues, C. Studievic, and P. Maupas, J. Clin. Pathol., 23, 540-544 (1970).

869. J. A. Varley, Analyst (London), 91, 119-126 (1966).

870. J. A. Varley and K. F. Baker, Analyst (London), 96, 734-738 (1971).

871. P. Vestergaard and J. F. Sayegh, in Advances in Automated Analysis, Technicon International Congress 1969, Vol. I, Mediad, Inc., White Plains, N.Y., 1970, pp. 327-332.

872. P. Vestergaarde and S. Vedso, J. Chromatogr., 19, 512-521 (1965).

873. J. K. Viktora and A. Baukal, in Automation in Analytical Chemistry, Technicon Symposia 1967, Vol. I, Mediad, Inc., White Plains, N.Y., 1968, pp. 447-449.

874. K. Von Berlepsch, Anal. Biochem., 27, 424-432 (1969).

875. G. Voss, J. Ass. Offic. Anal. Chem., 52, 1027-1034 (1969).

876. J. H. Vought, in Advances in Automated Analysis, Technicon International Congress 1969, Vol. II, Mediad, Inc., White Plains, N.Y., 1970, pp. 305-308.

877. G. G. Vurek, Anal. Biochem., 15, 171-175 (1966).

878. J. L. Wachtel and P. W. Peterson, in Automation in Analytical Chemistry, Technicon Symposia 1965, Mediad, Inc., New York, 1966, pp. 3-6.

879. H. E. Wade and B. P. Phillips, Anal. Biochem., 44, 189-199 (1971).

880. W. H. Wagner, in Automation in Analytical Chemistry, Technicon Symposia 1965, Mediad, Inc., New York, 1966, pp. 160-163.

881. B. J. Wahl and G. Auger, in Advances in Automated Analysis, Technicon International Congress 1969, Vol. II, Mediad, Inc., White Plains, N.Y., 1970, pp. 273-281.

882. L. Walker and E. Amador, Clin. Chem., 18, 568-570 (1972).

883. W. H. C. Walker, Clin. Chim. Acta, 32, 305-306 (1971).

883a. W. H. C. Walker and K. R. Andrew, Clin. Chim. Acta, 57, 181-185 (1974).

884. W. H. C. Walker, C. A. Pennock, and G. K. McGowan, Clin. Chim. Acta, 27, 421-435 (1970).

885. W. H. C. Walker, J. C. Shepherdson, and G. K. McGowan, Clin. Chim. Acta, 35, 455-460 (1971).

886. W. H. C. Walker, J. Townsend, and P. M. Keane, Clin. Chim. Acta, 36, 119-125 (1972).

887. V. Wallace, Anal. Biochem., 20, 411-418 (1967).

888. V. Wallace, Anal. Biochem., 20, 517-524 (1967).

889. K. A. Walsh, R. M. McDonald, and R. A. Bradshaw, Anal. Biochem., 35, 193-202 (1970).

890. A. M. Ward and A. D. Hirst, Amer. J. Clin. Pathol., 51, 751-759 (1969).

891. M. H. Warner and J. B. Jones, in Automation in Analytical Chemistry, Technicon Symposia 1966, Vol. I, Mediad, Inc., White Plains, N.Y., 1967, pp. 145-148.

892. A. M. Webb, Clin. Chem., 15, 764 (1969).

893. C. E. Webber, J. H. Johnstone, and E. S. Garnett, Clin. Chem., 15, 219-223 (1969).

894. D. J. Weber, H. Mitchner, L. Kaldy, and K. Yasaki, J. Pharm. Sci., 60, 467-468 (1971).

895. P. Weber, I. Bornstein, and R. J. Winzler, Anal. Biochem., 14, 100-105 (1966).

896. L. H. Weinstein, R. F. Bozarth, R. H. Mandl, S. Gross, and J. F. Goldman, Anal. Biochem., 11, 155-158 (1965).

897. M. Werner, in Automation and Data Processing in the Clinical Laboratory (G. M. Brittin and M. Werner, eds.), Thomas, Springfield, Ill., 1970, pp. 47-56.

898. W. Weschler, S. Allen, and K. Negersmith, in Advances in Automated Analysis, Technicon International Congress 1970, Vol. I, Thurman Associates, Miami, 1971, pp. 431-436.

899. G. Westlake, D. K. McKay, P. Surh, and D. Seligson, Clin. Chem., 15, 600-610 (1969).

900. L. G. Whitby, Brit. Med. J., 1964 (2), 895-899 (1964).

901. L. G. Whitby, F. L. Mitchell, and D. W. Moss, in Advances in Clinical Chemistry (O. Bodansky and C. P. Stewart, eds.), Vol. 10, Academic Press, New York, 1967, pp. 65-156.

902. L. M. White and M. A. Gauger, Anal. Biochem., 23, 355-357 (1968).

903. L. M. White and M. A. Gauger, in Automation in Analytical Chemistry, Technicon Symposia 1967, Vol. I, Mediad, Inc., White Plains, N.Y., 1968, pp. 213-217.

904. E. C. Whitehead, in Automation in Analytical Chemistry, Technicon Symposia 1965, Mediad, Inc., New York, 1966, pp. 437-451.

905. E. C. Whitehead, in Automation in Analytical Chemistry, Technicon Symposia 1966, Vol. I, Mediad, Inc., White Plains, N.Y., 1967, pp. 364-367.

906. E. C. Whitehead, Proc. 1963 Technicon Symposium, Automated Anal. Chem. (London), pp. 45-68.

907. R. Whitehead, G. H. Cooke, and B. T. Chapman, in Automation in Analytical Chemistry, Technicon Symposia 1967, Vol. II, Mediad, Inc., White Plains, N.Y., 1968, pp. 377-380.

908. R. W. Whitley and H. E. Alburn, Ann. N.Y. Acad. Sci., 130, 634-646 (1965).

909. R. W. Whitley and H. E. Alburn, 1964 Technicon International Symposium, Interim Reprint, New York, "Semi-Automated Method for the Determination of Serum Phospholipids."

910. P. Wilding, Clin. Chim. Acta, 8, 918-924 (1963).

911. E. B. Williams and R. B. Lyons, Anal. Biochem., 42, 342-349 (1971).

912. M. R. Wills and B. C. Gray, J. Clin. Pathol., 17, 687-689 (1964).

913. A. L. Wilson, Analyst (London), 90, 270-277 (1965).

914. A. M. Wilson, Anal. Chem., 38, 1784-1786 (1966).

915. B. W. Wilson, Clin. Chem., 12, 360-367 (1966).

916. G. D. Winter, Ann. N.Y. Acad. Sci., 87, 875-882 (1960).

917. D. J. Wood and A. R. Cousins, in Automation in Analytical Chemistry, Technicon Symposia 1966, Vol. II, Mediad, Inc., White Plains, N.Y., 1967, pp. 97-102.

918. W. A. Wood and S. R. Gilford, Anal. Biochem., 2, 589-600 (1961).

919. K. Wrightman, in Advances in Automated Analysis, Technicon International Congress 1969, Vol. II, Mediad, Inc., White Plains, N.Y., 1970, pp. 61-67.

920. K. B. Wrightman, R. E. Barber, and R. F. McCadden, in Automation in Analytical Chemistry, Technicon Symposia 1965, Mediad, Inc., New York, 1966, pp. 12-21.

921. K. B. Wrightman and W. W. Holl, Ann. N.Y. Acad. Sci., 130, 516-524 (1965).

922. K. B. Wrightman and R. F. McCadden, Ann. N.Y. Acad. Sci., 130, 827-834 (1965).

923. T. Wurzburger, in Advances in Automated Analysis, Technicon International Congress 1970, Vol. II, Thurman Associates, Miami, 1971, pp. 71-73.

924. H. Y. Yee and A. Zin, Clin. Chem., 17, 950-953 (1971).

925. D. S. Young, J. Clin. Pathol., 19, 397-399 (1966).

926. D. S. Young, R. M. Montague, and R. R. Snider, Clin. Chem., 14, 993-1001 (1968).

927. W. R. Young, in Advances in Automated Analysis, Technicon International Congress 1970, Vol. II, Thurman Associates, Miami, 1971, pp. 259-266.

928. R. S. Yunghans and W. A. Munroe, in Automation in Analytical Chemistry, Technicon Symposia 1965, Mediad, Inc., New York, 1966, pp. 279-284.

929. J. B. Zagar, P. P. Ascione, and G. P. Chrekian, J. Ass. Offic. Anal. Chem., 54, 1272-1276 (1971).

930. J. S. Zajac, Amer. J. Clin. Pathol., 45, 651-652 (1966).

931. J. S. Zajac, in Automation in Analytical Chemistry, Technicon Symposia 1966, Vol. I, Mediad, Inc., White Plains, N.Y., 1967, pp. 240-245.

932. B. Zak and E. S. Baginski, Anal. Chem., 34, 257-259 (1962).

933. B. Zak and E. Baginski, Chemist-Analyst, 51, 39-40, 42 (1962).

934. B. Zak and E. Epstein, Clin. Chem., 11, 641-644 (1965).

935. B. Zak, J. Holland, and L. A. Williams, Clin. Chem., 8, 530-537 (1962).

936. R. A. Zaroda, Amer. J. Clin. Pathol., 41, 377-380 (1964).

937. G. H. Zeman, P. Z. Sobocinski, and R. L. Chaput, Anal. Biochem., 52, 63-68 (1973).

938. R. C. Zerfing and H. Veening, Anal. Chem., 38, 1312-1316 (1966).

939. R. C. Zerfing and H. Veening, in Automation in Analytical Chemistry, Technicon Symposia 1966, Vol. I, Mediad, Inc., White Plains, N.Y., 1967, pp. 435-439.

940. J. A. Zivin and J. F. Snarr, Anal. Biochem., 52, 456-461 (1973).

941. M. M. Zuckerman and A. H. Molof, in Advances in Automated Analysis, Technicon International Congress 1969, Vol. II, Mediad, Inc., White Plains, N.Y., 1970, pp. 121-124.

AUTHOR INDEX

Numbers in parentheses are reference numbers and indicate that an author's work is referred to although his name is not cited in the text. Underlined numbers give the page on which the complete reference is listed.

A

Abdullah, M. I., 21, 187, 192, 195(1), 199(1), <u>227</u>

Acevedo, H. F., 41(803), <u>273</u>

Adelman, M. H., 48, 123, 123(4), 130(4), 136, <u>227</u>

Adler, H. J., 43(245, 246), 48(245, 246), 168(251), 175(245, 246), 179(245, 246), 181(245, 246), <u>241</u>

Adler, S. L., 37(656), <u>265</u>

Adler, S. L., Sr., 4, <u>227</u>

Afghan, B. K., 67, 92, 100, 124n(6), 152, 152(339), 156(339), <u>227</u>, <u>247</u>

Agren, A., 25, <u>227</u>

Ahuja, J. N., 74, <u>227</u>

Ahuja, S., 35, 36, 86, 86(11), 87, 87(12), 113, 142, 146, 162(12), <u>227</u>, <u>228</u>

Alber, L. L., 42, 113, 189, 190, 192, 195(14), 201, <u>228</u>

Albisser, A. M., 148, <u>228</u>

Albright, B. E., 150, <u>228</u>

Alburn, H. E., 13(908), 14(909), 32, 33, 61(908, 909), 61n(908, 909), <u>280</u>

Allen, E., 32, 33, 58, <u>228</u>

Allen, S., 84(898), <u>279</u>

Alpert, N. L., 5(19), <u>228</u>

Alsos, I., 186, 195, 196(21), 197(21), <u>228</u>

Amador, E., 19, 20(882), 27(24), 59, 65(24), 73, 102, 114, 127, 167, <u>228</u>, <u>278</u>

Amako, H., 177(436), <u>252</u>

Ambrose, J. A., 27, 40, 40(27), 41, 45, 66(25, 27), 73(25, 27), 162, <u>228</u>, <u>229</u>

Anderson, N. G., 112, <u>229</u>

Anderson, R. A., 45(31, 32), 110, 115, 142, 143, <u>229</u>

Anderson, R. C., 9(41), <u>229</u>

Andrew, K. R., 211(883a), 223(883a), <u>278</u>

Angellis, D., 163(279), 165(279), <u>243</u>

Annino, J. S., 28, <u>229</u>

Annokkee, Y., 124n(197), 171, <u>238</u>

Anselm, C. D., 168, <u>248</u>

Anstiss, C. L., 165, <u>229</u>

Antonis, A., 59, 61, 63, 64, 70, 78, 103, <u>229</u>

Appleton, H. D., 38(600), <u>262</u>

Arbogast, J. L., 42, <u>229</u>

Archer, N. P., 9(41), <u>229</u>

M

SUBJECT INDEX

A

Acetonitrile, transmission tubing for, 164

Acetylkynurenine, determination of, 155

Acid phosphatase, determination of, 163

Acids:
  buffer-indicator acid-base determinations of, 171
  carboxylic, chromatography of, 175
  free, determination in presence of hydrolyzable ions, 149
  normality of, in uranyl nitrate solution, 163

Adenosine triphosphate, determination of, 108, 111

Alcohol:
  determination of, 183, 184, 206
  by gas chromatography, 105
  distillation and determination of, 157

Alcohols:
  chromatography of, 175
  fatty, determination of, 37

Aldehydes:
  atmospheric, determination of, 168
  volatile, distillation of, 157

Alkaline phosphatase:
  chromatography of, 180
  determination of, 169, 180
  reaction kinetics of, 160

Alkaloids, steam distillation and determination of, 111, 154

Aluminum:
  determination of, 18, 31
    by graphite boat, induction furnace, and atomic absorption spectrophotometer, 93

Amides, hydrolysis rates of, 167

Amines:
  adsorption on tubing, 165
  aromatic, steam distillation and determination of, 154
  extraction and determination by "dye complex" method, 140, 147

o-Aminoacetophenone, steam distillation and determination of, 155

Amino Acid Analyzer (Technicon):
  fail-safe devices, 181
  uses of, review, 173

Amino acids, chromatography of, 175, 180

α-Amino acids, in column eluates, monitoring by polarography, 100, 150, 182

Amino sugars, chromatography of, 175

Printed and bound by CPI Group (UK) Ltd, Croydon, CR0 4YY

23/10/2024

01778224-0013